Mathematics and the Body

This book explores alternative ways to consider the relationship between mathematics and the material world. Drawing on the philosophy of Gilles Châtelet and the posthumanist materialism of Karen Barad, the authors present an 'inclusive materialist' approach to studying mathematics education. This approach offers a fresh perspective on human and non-human bodies, challenging current assumptions about the role of the senses, language and ability in teaching and learning mathematics. Each chapter provides empirical examples from the classroom that demonstrate how inclusive materialism can be applied to a wide range of concerns in the field. The authors analyse recent studies on students' gestures, expressions and drawings in order to establish a link between mathematical activity and mathematical concepts. *Mathematics and the Body* expands the landscape of research in mathematics education and will be an essential resource for teachers, students and researchers alike.

Elizabeth de Freitas is an associate professor at the Ruth S. Ammon School of Education at Adelphi University. She is the co-editor of *Opening the Research Text: Critical Insights and In(ter)ventions into Mathematics Education* (2008) and an associate editor of the journal *Educational Studies in Mathematics*.

Nathalie Sinclair is an associate professor in the Faculty of Education, an associate member in the Department of Mathematics and a Canada Research Chair in Tangible Mathematics Learning at Simon Fraser University. She is also an associate editor of *For the Learning of Mathematics*. She is the author of *Mathematics and Beauty: Aesthetic Approaches to Teaching Children* (2006) and *Developing Essential Understanding of Geometry for Teaching Mathematics* (2012), among other books.

LEARNING IN DOING: SOCIAL, COGNITIVE AND COMPUTATIONAL PERSPECTIVES

SERIES EDITOR EMERITUS

John Seely Brown, *Xerox Palo Alto Research Center*

GENERAL EDITORS

Roy Pea, *Professor of Education and the Learning Sciences and Director, Stanford Center for Innovations in Learning, Stanford University*

Christian Heath, *The Management Centre, King's College, London*

Lucy A. Suchman, *Centre for Science Studies and Department of Sociology, Lancaster University, UK*

BOOKS IN THE SERIES

The Construction Zone: Working for Cognitive Change in School
Denis Newman, Peg Griffin, and Michael Cole

Situated Learning: Legitimate Peripheral Participation
Jean Lave and Etienne Wenger

Street Mathematics and School Mathematics
Terezinha Nunes, David William Carraher, and Analucia Dias Schliemann

Understanding Practice: Perspectives on Activity and Context
Seth Chaiklin and Jean Lave, Editors

Distributed Cognitions: Psychological and Educational Considerations
Gavriel Salomon, Editor

The Computer as Medium
Peter Bøgh Anderson, Berit Holmqvist, and Jens F. Jensen, Editors

Sociocultural Studies of Mind
James V. Wertsch, Pablo del Rio, and Amelia Alvarez, Editors

Sociocultural Psychology: Theory and Practice of Doing and Knowing
Laura Martin, Katherine Nelson, and Ethel Tobach, Editors

(Continued after index)

Mathematics and the Body

Material Entanglements in the Classroom

ELIZABETH DE FREITAS
Adelphi University

NATHALIE SINCLAIR
Simon Fraser University

CAMBRIDGE
UNIVERSITY PRESS

CAMBRIDGE
UNIVERSITY PRESS

32 Avenue of the Americas, New York, NY 10013-2473, USA

Cambridge University Press is part of the University of Cambridge.

It furthers the University's mission by disseminating knowledge in the pursuit of education, learning and research at the highest international levels of excellence.

www.cambridge.org
Information on this title: www.cambridge.org/9781107039483

First published 2014

Printed in the United States of America

A catalogue record for this publication is available from the British Library.

Library of Congress Cataloguing in Publication data
De Freitas, Elizabeth.
Mathematics and the body : material entanglements in the classroom / Elizabeth de Freitas, Adelphi University, Nathalie Sinclair, Simon Fraser University.
pages cm. – (Learning in doing. Social, cognitive and computational perspectives)
Includes bibliographical references and index.
ISBN 978-1-107-03948-3 (hardback)
1. Mathematics – Study and teaching. 2. Mathematics – Philosophy.
I. Sinclair, Nathalie. II. Title.
QA9.D27 2014
510.71–dc23 2014002452

ISBN 978-1-107-03948-3 Hardback

Contents

Figures

Series Foreword

This series for Cambridge University Press is widely known as an international forum for studies of situated learning and cognition. Innovative contributions are being made by anthropology; by cognitive, developmental and cultural psychology; by computer science; by education; and by social theory. These contributions are providing the basis for new ways of understanding the social, historical and contextual nature of learning, thinking and practice that emerges from human activity. The empirical settings of these research inquiries range from the classroom to the workplace, to the high-technology office, and to learning in the streets and in other communities of practice. The situated nature of learning and remembering through activity is a central fact. It may appear obvious that human minds develop in social situations and extend their sphere of activity and communicative competencies. But cognitive theories of knowledge representation and learning alone have not provided sufficient insight into these relationships. This series was born of the conviction that new, exciting interdisciplinary syntheses are underway, as scholars and practitioners from diverse fields seek to develop theory and empirical investigations adequate for characterizing the complex relations between social and mental life and for understanding successful learning wherever it occurs. The series invites contributions that advance our understanding of these seminal issues.

Roy Pea
Christian Heath
Lucy A. Suchman

Foreword

Brian Rotman

Elizabeth de Freitas and Nathalie Sinclair have written an admirable and provocative book. Ambitious, original and theoretically accomplished, its purpose is to develop a new materialist approach, what they call 'inclusive materialism', to the learning of mathematics – one that includes and foregrounds the activity of the body against the long-standing mentalist conception of mathematics as an activity of pure, abstract thought. Extending the current turn to materialism in philosophy and the humanities to mathematics, they reject Kantian-based epistemological schemes that understand knowledge as perception filtered through internal, *a priori* intuitions and the conceptual categories, in favour of a more Humean, empiricist approach that gives primacy to external sensation; ontologically, they reject Platonic realism, the belief that mathematical objects – points, numbers, lines and so on – are immaterial entities that exist in some Platonic heaven – 'out there', beyond time, space, matter – while mathematical activity consists of discovering truths about those objects, which is analogous to scientists studying external reality. Despite numerous critiques – the chief of which asks how material beings can make contact with things in a transcendent heaven – this metaphysical idealism remains the conventional belief, defended and widely embraced by mathematicians and others.

Mathematics and the Body is directed to mathematics educators and validates, as well as explicates, its ideas by critically examining a series of experimental classroom lessons designed by the authors and by others, which focus on fundamental mathematical concepts such as number, parallelism, circles and diagrams. As the authors observe, the issue of embodied mathematics in education is topical. In the last decade or so, a growing number of differently oriented initiatives – cognitive, phenomenological, enactive, communication-based approaches – have been devoted to examining the role played by students' bodies: their gestures; hand, eye and limb movements; their verbalizations; their drawings and diagrams; and their relation to the tokens, devices, physical objects and surfaces with which they interact. The book aims to explore the assumptions and consequences of this work. To do so, and to go beyond it, they pose and confront the fundamental question: 'How are the physical aspects of mathematical activity – be it that of students or mathematicians – transformed into the so-called abstractions and generalisations of formal mathematics?' Their answer involves formulating a new, extended notion of 'body' and correlatively a material understanding of the mathematical concepts with which such a body engages. The inclusion of mathematicians' physical activity in their question indicates a possible parallel between the creation of mathematics and its re-creation by students in the classroom – a link, that is, between the history of mathematics and the learning of it. Such is indeed the case, as is evident in their opening sentence: 'The idea for this book began as we read Gilles Châtelet's (1993/2000) stunning book on the history of mathematics, which challenges many long-standing, as well as contemporary, philosophies of mathematics.' Châtelet's book, *Figuring Space*, opens up several key moments in the historical development of the subject, demonstrating how the interrelation of gesture – resulting from 'disciplined movements of a body' – and physical diagrams operate at the heart of mathematical invention. De Freitas and Sinclair embrace Châtelet's linking of gestural bodies and formal abstractions and work to import it into the mathematics classroom.

But before they can accomplish this, they need to establish the nature of embodiment. 'When', they ask, 'does a body become a body?' A survey of the mathematical embodiment literature finds them critical of approaches that fail to escape the 'dualistic tradition of the mind/body split'; or that 'demote the body to acting merely as the vessel or container of some higher act of cognition'; or that 'centre human will or intention in the orchestrating of experience', assuming the human body to be 'the principal administrator of its own participation'. Moreover, locating knowing and agency in the individual body does not adequately address the collective social body. Where, then, are the boundaries of a body? Against the commonsense view that 'the body is an individual, discrete entity and that cognition occurs within its borders', the authors turn to posthumanist discourses of subjectivity and agency, according to which subjects are dynamic assemblages of dispersed social networks, and the 'human body itself must be conceived in terms of malleable borders and distributed networks'; that is, a body understood as a 'set of material relations that seems to structure the other material relations around it'. In the classroom, as they illustrate in their analyses of students' activities, such an assemblage-body will be composed of 'humans, writing implements, writing surfaces, texts, desks, doors, chips, as well as disciplinary forces and habits of control and capitulation'. A consequence of conceiving the body in this way is that agency and thought become distributed across multiple sources in the students' physical and psycho-social environment. Thus, analogous to Nietzsche's insistence on 'deeds without a doer', one can have 'thoughts without a thinker', in the sense that the source of thought can come from material relations outside or beside the isolated thinking self, a phenomenon that Gilles Deleuze, whose materialist ideas exert a profound effect on the authors' project, calls the 'exteriority of thought'. In short, the power and efficacy of a body in relation to mathematics must be understood as distributed across an assemblage of heterogeneous relations, a posthumanist understanding not to be identified with the capacity that is 'localized in a human body or in a collective produced (only) by human efforts'.

But how does this material body-assemblage become entangled with mathematical concepts? In what sense can we consider concepts, mathematical or otherwise, to be related to *matter*? The question goes to the theoretical heart of *Mathematics and the Body*. The authors' aim is to show how 'mathematical concepts partake of the material in operative, agential ways'. In order to accomplish this, they need to go outside a humanist conception of matter and 'materiality', as well as construct a new approach to the nature of concepts. They derive this from contemporary feminist philosophers, principally Karen Barad, but also Jane Bennett, and Diana Coole and Samantha Frost, whose common aim is to reorient how we think about 'matter' and the material world. From Barad's theory of 'agential realism', derived from Niels Bohr's explication of quantum phenomena, they take the understanding of a concept not as an immaterial mental object, but as 'a material arrangement of things' and of relations preceding, and in some sense constituting, that to which they relate, so that things are always 'intra-related', rather than interrelated. Jane Bennett's concept of 'vibrant matter', a (non-animistic) understanding that credits matter with agency, and the 'inclusive materialism' of Coole and Frost provide the wherewithall for constructing a body-concept nexus. This, along with the anthropological work of Lambros Malfouris and Bruno Latour, allows them to rethink the concept of 'mere' matter. Rejecting the Cartesian split between the active, cognizing human mind and inert, 'dead' matter – the contemporary orthodoxy underpinning the physical sciences' engagement with matter – these various thinkers urge materialisms in which the freedom and agency that Descartes restricted to the embodied human mind is opened up and dispersed across human and non-human agents. The ontology of mathematics that the authors weave from these diverse materialisms, with their insistence on the extra-human and material dimensions of thought, complements the authors' construction of the assemblage-body. With this theoretical meshing in place, de Freitas and Sinclair are ready to expand on how the two – bodies and formal mathematical concepts – might in practice become entangled. They accomplish this through the essay of Châtelet that inspired them to pursue their ambitious body-mathematics project.

Châtelet's interest is in how mathematics comes into being – its genesis, its becoming rather than 'being' – and his essay is a series of analyses of specific mathematical inventions, such as Grassmann's creation of algebras over vector fields and Cauchy's method of integrating complex functions, that reveal the physico-conceptual movements that constitute them.

His starting point is actual, physical movement. According to Châtelet, the 'amplifying abstractions' of mathematics, whatever their ultimate immaterial representation as formal constructs may be, have bodily beginnings. They originate in gestures, 'disciplined distributions of mobility', that are not signs or representations of anything prior to or outside themselves, but instead are material events that, through their actions and by the fact of their occurrence, bring new mathematical meanings into being. They are not, Châtelet insists, describable by formal languages, cannot be determined by algorithms, are not expressions of an intention (although they can be retrospectively seen as such), and are not in fact consciously produced: 'One is', he says, ' infused with the gesture before knowing it.' And they do not work through reference or signification, but rather by pointing, through allusions that – in interaction with diagrams (which are themselves responses to problems) – give rise to 'dynasties of problems' and correlative families of ever more precise allusions. A diagram, for Châtelet, is a frozen gesture, a gesture caught mid-flight in its path towards a formal abstraction: It can 'transfix a gesture, bring it to rest, long before it curls up into a sign'. Diagrams are intermediaries between bodies and mathematical objects and operations. They are, like gestures, material events. Contrary to the customary view of them, they are not depictions, illustrations or visual icons of mathematical objects or concepts (although they can be), but instead are pivotal devices in the creation of mathematical meaning – 'kinematic capturing devices', as the authors neatly describe them, 'for direct sampling that cut up space and allude to new dimensions and new structures'.

In a sense, diagrams are works in progress, never complete in themselves: '[I]f [a diagram] immobilizes a gesture in order to set down an operation, it does so by sketching a gesture that then cuts out another.' Diagrams and gestures interact, mutually presupposing each other, participating in what the authors call each other's 'provisional ontology'. Overall, the gesture-diagram nexus operates as a 'dynamic process of excavation that conjures the sensible in sensible matter'. The authors relate this conjuring to Barad's realist understanding of concepts as material arrangements. 'The concept itself', they observe, 'is entailed in the hands that gesture, the mouth that speaks, and the affect that circulates across an interaction.' They concretize this entailment through a variety of examples that range from discussing how 'the point at infinity' is cognized in projective geometry to describing at length the results of an experiment with a class of undergraduates asked to draw diagrams in response to a simple film of moving circles.

The gesture-diagram apparatus of allusions to mathematical meanings is one-half of what the authors find valuable in Châtelet's approach; the other is his deployment of the notion of the *virtual*. He takes this from Deleuze's materialist and immanentist philosophy, according to which the physical world of matter constantly comes into being – becomes – by making actual that which is virtual: 'The virtual must be defined as strictly a part of the real object – as though the object had one part of itself in the virtual into which it is plunged as though into an objective dimension.' The virtual is that which is latent in matter, the source of all that it could become, which the authors interpret as its 'mobility, vibration, potentiality and indeterminacy', and it is the link Châtelet provides between the mathematical and physical worlds.

Following Gottfried Leibniz in conceiving space as 'a flexible, folding and animated substance', Châtelet observes that the supposedly immovable objects of mathematics divorced from 'sensible matter' are, on the contrary, always in a state of potential movement and change; a geometrical point (line, circle) cannot be confined to a designated entity, the representation of a position within a fixed, absolute space. As he observes in the case of Cauchy's treatment of a singular point in the complex plane, the virtuality of a point, probed by mathematicians within 'thought-experiments', becomes the source of radically new concepts. A point is the simplest example of a diagram, but the effect is quite general. As the authors observe, 'the virtual or potentiality of a[ny] diagram consists of all the gestures and future alterations that are in some fashion "contained" in it'. Mathematical entities, then, are material objects with virtual and actual dimensions. The virtual is not so much a bridge – an interrelation between mathematics and the physical world, as if they were initially separate and then joined – as an 'intra-relation' – which Barad defines as a mutual fabrication or co-constitution, wherein the two are thoroughly entangled. This means that mathematical concepts engage in a process of becoming which binds them to the actions of mathematicians, leading to the authors' striking conclusion that '[t]he mathematical body comes into being through actualizing the virtual – through gestures, diagrams and digital networks, *we become mathematics*; we incorporate and are incorporated by mathematics" (emphasis in original).

Summarized in this way and taken in isolation, the concept of 'becoming' mathematics will doubtless strike many potential readers of *Mathematics and the Body* as a strange and counter-intuitive characterization of 'we' and of mathematics, but hopefully this will not be their experience. Throughout, de Freitas and Sinclair seem fully aware of the unfamiliarity of the ideas they mobilize and of the conceptual demands of their thesis; they go to considerable lengths to present matters as accessibly as possible. Not only does their book carefully develop the ideas of the body-as-assemblage and the body's dynamic relation with abstract concepts that forms the basis of how we become mathematics, but it also contains a wealth of material and a rich texture of connections that elaborate and contextualize their thesis. Thus, beside constantly rooting their ideas in the concrete classroom observations and experiments which feature throughout, they step back and offer a series of illuminating and provocative chapter-length discussions of key aspects of their field, ranging from the 'sensory politics of the body mathematical' and 'mapping the mathematical aesthetic' to the 'materiality of language' and the material dimension of 'inventiveness in the mathematics classroom'.

In a final reflection on what becoming mathematics might mean, both generally and in the context of the classroom, they invoke Deleuze's concept of a 'minor science', a 'minor literature', and indeed a 'minor mathematics' – forms of thought and creation which escape the constrictions of the dominant 'state', or orthodox version. They describe 'a mathematics that is not the state-sanctioned discourse of school mathematics, but that might be full of surprises, non-sense and paradox' and which, although at odds with institutional demands and the domination of a fixed curriculum, 'is likely to engage students and teachers in more expansive ways, and [their] hope is that it would engage *more* students in mathematics'. Whether or not it does remains, of course, to be seen, but in any event the minor mathematics that Elizabeth de Freitas and Nathalie Sinclair usher onto the mathematics education scene constitutes a major theoretical intervention in their field. *Mathematics and the Body* is a valuable, radical and challenging work.

Acknowledgements

We are deeply grateful to those friends and colleagues who spurred us to write this book. Their words of support and challenge were crucial in developing our ideas and in helping us weave together many different theoretical threads. We set out to write an exploratory book about mathematics and the body and were instantly humbled by the incredible amount of insightful work already in progress on this topic. We want to thank Tony Brown in particular for pressing us to continue with our own vision and commitment to philosophical concerns in this area. The 'Contemporary Theories in Mathematics Education' conference in Manchester in 2011, which was jointly organised by Tony, Yvette Solomon and Julian Williams, first ignited our work together and gave us the opportunity to present some of the main ideas found in Chapter 3. We would also like to thank Ricardo Nemirovsky for his detailed and careful reading of the manuscript and his exceptional ideas for how to better frame questions about the body. We benefitted a great deal from our visit to UCSD, during which, through lectures and discussions, we were able to work out many wrinkles that emerged in the writing of Chapters 1 and 2. We are tremendously in debt to David Pimm for his excellent suggestions on how to revise various arguments throughout the book, his ongoing enthusiasm for the project, and his help with both mathematical and historical details. We are also grateful to those who read chapters throughout the process and who offered their time in submitting lots of astute and discerning comments, in particular Alf Coles, Beth Herbel-Eisenmann, Francesca Ferrara, Michael Harris, Lulu Healy and Margaret Walshaw. Some of the ideas that we develop in this book were initially explored in articles published in *Educational Studies in Mathematics*, *ZDM*, *Educational Philosophy and Theory*, and the *Journal of Humanistic Mathematics*, and we are grateful to the reviewers and editors of these publications for that opportunity. Two names that come up a number of times throughout the book, attached to people who generously reviewed and corresponded about this work, are Luis Radford and Michael-Wolff Roth. We would also like to thank Oi-Lam Ng for her valuable assistance in preparing the manuscript. Finally, we are very grateful to Cambridge University Press for its support in bringing our writing to publication; we are particularly grateful to Kori Lisa Yee Litt of PETT Fox, Inc., for outstanding editorial work.

Nathalie wishes to also acknowledge the support she received from the Canada Research Chair program, which enabled her to devote long, continuous hours to reading and writing. She would also like to thank Sean Chorney for many engaging conversations on the topics of agency, new materialism and posthumanism. Elizabeth would like to thank Paola Valero for organising the body-course in Denmark, and also Hillevi Lenz Taguchi, Anna Palmer, Lisa Mazzei and Lisa Björklund Boistrup for organising various stimulating conversations about materiality, virtuality and concepts at the Stockholm Event. She would also like to thank Jerry Rosiek for the enjoyable speculative conversations about theory, as well as her friends in the Deleuze Reading Group in Brooklyn.

Introduction

The idea for this book began as we read Gilles Châtelet's (1993/2000) stunning book on the history of mathematics, which challenges many long-standing, as well as contemporary, philosophies of mathematics. His book places gestures and diagrams at the centre of mathematical inventiveness, which struck many chords for us both, not least in relation to our mutual interest in the role of these body-based and mobile devices in the teaching and learning of mathematics. We saw in Châtelet a way of better understanding how materiality might matter for mathematics, which has for so long been taken as an abstract and static discipline that resists any links with the physical world. Although we have learned a great deal from recent scholars working on the embodied nature of mathematical thinking and learning, we were unsatisfied with some of their basic philosophical assumptions about the nature of both mathematics and the body. Further, we wanted to find ways of being able to work with an embodied mathematics, while also drawing on the powerful and insightful research of colleagues pursuing a more discursive and politicised way of understanding the teaching and learning of mathematics. This latter research usefully moves beyond the essentialist assumptions of acquisitionist theories of learning and sheds considerable light on the sociocultural facets of education, but it has frequently neglected the role of the body in teaching and learning. Châtelet inspires a new kind of materialist study of mathematics and the body, allowing for new ways of exploring how mathematics partakes of the material world. In this book, we try to show how this new approach can be put to work in rethinking mathematics education.

As we read Châtelet, we began to see links with the contemporary work of 'new' materialists such as Karen Barad, as well as philosophers such as Gilles Deleuze and Brian Rotman. Their writing enabled us to see how we could talk about embodiment *and* discourse in coherent ways. They linked us to important traditions in philosophy, feminism, history and mathematics that enabled us to contextualize and question work in mathematics education. They also gently but steadily led us into rethinking the politics and aesthetics of mathematics – issues that we had both pursued independently before and to which we were eager to return. While our own starting point was Châtelet, we begin this book by pushing on two very central, and perhaps counter-intuitive, notions that we see as underlying Châtelet's philosophy. The first is the very notion of the human body, as we unpack what it is and what it might be. The second is the nature of materiality and its relation to the human body, the social and the conceptual. Our work on these notions (body and matter), particularly in the context of mathematics education, has led us to a particular brand of materialism that we call *inclusive materialism*.

In Chapter 1, we propose a rethinking of the assumed boundaries of the body and the taken-for-granted geography of the body's interaction with the material world. Our impetus for doing so comes from questions that arose for us as we have made our way through the vast amounts of work on the body and its relation to learning. What is a body? Where does it begin and end? When does individuation of a body occur in a classroom? Must there be an interior and exterior? Is the body a bounded organism? These are questions that became more pressing as we tried to absorb the implications of Châtelet's materialist philosophy of mathematics, which refuses to see the body – any body, not just the human body – as fixed, stable or unitary. Indeed, how can we continue to speak of a bounded body if, as Rotman (2008) suggests, the proliferation of our sensory powers through techno-societal developments in the last century have us 'becoming beside ourselves'? Many current theories of embodiment do not address these questions, perhaps because they are usually oriented towards analyses of student activity, trying to account for individual processes of learning. But if we trouble this enclosed 'definition' of the body, what might be the ramifications for what is meant by activity, by learning, by embodiment? One ramification will be the idea of the body as sometimes more and sometimes less than its physical parts, as inextricably bound up with artefacts, other bodies, concepts. Another will be the decentring of the human body, that is, the acknowledgement of non-human agencies also at play in any learning situation.

The second possibly counter-intuitive notion developed in Chapter 2 is the ontological entanglement of matter and meaning. Just as we often assume that the human body is contained by the contours of its skin, so we assume that matter is inert and entirely responsive to the will of human bodies. While some theorists have tried to accord agency to inanimate bodies as well, they have been mired in questions of intentionality – can a toaster *intend* to pop? – and much of the literature on the subject seems trapped in an ongoing dichotomizing of bodies, objects and actions. The notion of materiality that we have chosen to pursue, based largely on the work of Karen Barad, but also based on the concept of virtuality that one finds in the work of Gilles Deleuze, averts these problems by focusing on the indeterminacy of relations between various types of human and non-human agents. This allows us to extend materiality beyond the strict confines of concrete, physical objects so that meaning, discourse and concepts are also treated as material.

While Barad's materialism provides a compelling basis for rethinking meaning and matter, it stops short of being able to account for the seemingly abstract and immaterial nature of mathematics. This is where Châtelet comes in, providing a way of interpreting new materialist approaches, such as Barad's, in the case of mathematics. To be sure, his mathematics may not be a very familiar one because of the particular examples he selects, but also because he is proposing a radically different approach to the question of the ontological status of mathematical concepts. It certainly challenges the major philosophies of mathematics, but it might also cast mathematics in terms that are hard to reconcile with school mathematics. Nevertheless, this is the challenge we have taken on in this book, mapping this new approach onto school mathematics, situated within a post-humanist, materialist perspective that we call *inclusive materialism*. More generally, we approach mathematics from within an empiricist tradition by looking closely at the material specificities of mathematical experiences.[1] We pursue questions such as: What are the concrete material actions that constitute the activity of doing mathematics? What are the relations of exteriority – the relations between material parts – that comprise the corporeal habits of this cultural practice? Thus we position ourselves within a tradition in which abstract thought and materiality are assumed to be entwined. According to phenomenological currents within this tradition, thinking and reasoning – and

[1] The literature in mathematics education can be separated into two broadly conceived groups: the first abides by a Kantian-inspired theory of learning, in which it is argued that cognitive faculties synthesize sense perception; and the second aligns with a Humean-inspired approach, in which it is argued that conceptual categories are constituted through perceptual routine habits and material interactions. Unlike the Kantian tradition, which assumes that our experiences of the world are structured through internal categories or concepts that we impose on the material world of phenomena (Delanda, 2006), the Humean tradition is an empiricist one that lends itself to the study of emergent material habits and emergent cognitive structures.

any other related cognitive constructs – are always external or located in the 'flesh': 'Thinking is not a process that takes place "behind" or "underneath" bodily activity, but *is* the bodily activity itself' (Ferrara and Nemirovsky, 2005, p. 139). Given that empiricism comes in many different flavours, any claim must be seen as reflecting a particular set of cultural and scientific practices, practices that shift with history and context. This book problematizes any universal concept of the 'empirical' by drawing on historical arguments and emphasizing the ways in which research assumptions – both tacit and explicit – inform what we are able to see in our data.

We have not been exhaustive in this exploration; indeed, we have sometimes been opportunistic in choosing the particular aspects of mathematics education that we have worked on before, either individually or together. This is the case, for example, for our investigation of creativity and language use in the classroom, as well as for our re-examination of the mathematical aesthetic. At other times we have followed more closely some of the constructs that have been put forward by Barad and Châtelet, such as assemblages and virtuality, each of which is brought to bear on the concerns of mathematics education. One unexpected line of flight took us into the literature on disability, from which we learned a tremendous amount and in which we found kindred attempts to shift perspective along post-humanist materialist lines.

This book reflects out attempt to work through the implications of inclusive materialism in relation to mathematical activity. Our aim is to expand on Châtelet's sometimes poetic and difficult writing and to show through example and application how his insights are highly relevant to mathematics education. Much of the literature we draw on may be new to readers in mathematics education, but we hope that we will whet the appetite of those readers and perhaps spur others on in pursuing some of these ideas in their research. The idea of post-humanism that we develop is a difficult one to embrace – it seems to go against a common-sense inclination to centre the human subject in the study of teaching and learning. The idea of a body with unstable contours only provisionally individuated is also difficult to grasp given our life experiences and the seemingly definitive end to those experiences that occurs at the time of death. But we ask that the reader enter into a thought experiment – a 'what if' exercise – where such common-sense beliefs are set aside not only for the sake of thinking differently, but for the possibility of learning more about how mathematics and the human body consort. We offer this book as an invitation to explore what we deem to be fruitful ramifications from this particular way of thinking differently. Each chapter is meant to provoke and push the reader to consider alternative ways of conceptualizing the relationship between mathematics and matter. Thus the writing style is at times more evocative and poetic than what one usually finds in mathematics education literature. But we believe that this kind of speculative and creative work is extremely important in our field. Such work pushes the field into new uncharted terrain and allows for new conjectures about teaching and learning. Our aim in each chapter has been to show how ideas from new materialism can be put to work in opening up the landscape of research in mathematics education. Of course, no book can create an exhaustive map, and this has not been our objective. We have tried, however, to link our ideas with others who work in mathematics education, teasing out some of the at times subtle differences between various approaches. In each chapter, we pushed ourselves to perturb our own assumptions about the fundamental constructs that are often taken for granted in the field. We chose not to use just our own data throughout the book; instead, we discuss diverse kinds of examples that can be found in other publications and video resources, so that readers might access this material on their own.

Outline of the book

Chapter 1 is concerned with outlining the main philosophical assumptions and theoretical constructs about the body that are used in this book. We begin by examining the ways in which embodiment is currently conceptualized in mathematics education and the critiques that have been made of current conceptions of the body, both within mathematics education but also from without, by learning scientists, sociologists and post-humanist philosophers. We then pursue answers to some of the previously listed questions by proposing that the body need not be delimited by the container of its skin, nor should the concept of the body be exclusively reserved for sentient beings. We draw on feminist philosophy, and its own evolving quest to understand the nature of the female body, to better articulate how this new body might operate and to elaborate further on Rotman's sense of the body as that which is always becoming. This take on the body resonates powerfully with new materialism, which seeks to rethink the nature and role of matter and, in so doing, perturbs existing assumptions about how human beings interact with matter. Thus we argue that the body be conceived as an assemblage of diverse materialities in motion.

What is at stake in new materialism is ontology – what matter *is*. This is the focus of Chapter 2. The ontological questions we are posing are reminiscent of those posed by Warren McCullough (1965, p. 7) when he asked 'What is a number that Man may know it? And what is Man that he may know number?' But we try to shift the ground of this question so that knowing and becoming are more entangled. Chapter 1 provides the tools we need to see how McCullough's 'Man' and 'number' are involved in an ongoing dynamic such that it makes more sense to speak of man-number as a relationship than of either man or number as an independent, discrete entity. In other words, 'Man' and 'number' are part of a mutually constitutive material assemblage that has fewer boundaries and less fixity than McCullough's questions imply. Karen Barad's attempts to develop an approach to matter and concept in terms of *intra-activity* (a term she coins to avoid the assumption of distinct bodies acting with each other that the word interaction suggests) and an epistem-ontology are useful in this respect.

In Chapter 2, we build on the work of Barad to argue that theoretical concepts are inextricably material and that matter is intrinsically indeterminate. She elaborates her materialism by relying heavily on the work and writing of the physicist Niels Bohr. Indeed, she uses developments in twentieth-century physics to show how knowing and becoming are entangled. Although she does not write explicitly about mathematics or about learners, her focus is on the material nature of theoretical concepts, and thus we found her work highly pertinent to mathematics. But even if one acknowledges that concepts from physics – like electron, force or string – partake of the material world (hence making their materiality seem both natural and reasonable), it is likely harder to swallow the idea of the materiality of mathematical concepts. We can easily think of the way in which a concept like number *connects* or *applies* to the physical world, as in counting apples and measuring heights, but in what sense can it be said that number *partakes* of the material world? Theories of embodiment have argued that the sensorimotor experiences of humans – which are said to occur within the physical world – enable humans to create, understand and learn concepts like number. But such theories say next to nothing about what numbers are, let alone how mathematical concepts such as number can inhabit the physical world rather than some metaphysical, Platonic realm. We suspect that some of the research on embodiment may in fact serve to entrench further an image of mathematical concepts as universal, static and pregiven, despite the new focus on the role of the body in learning, in part because the mathematics is not adequately historicized. Our hope is that the new kinds of materialisms that we explore in this book will supply us with a set of theoretical tools to study more thoroughly mathematical concepts as partaking of the physical world.

In Chapter 3, we extend the discussion of the materiality of mathematical concepts with the help of Châtelet's work on the role of gestures and diagrams in mathematical invention. Gestures and diagrams both have received increased attention in mathematics education, especially within the recent emergence of embodied and semiotic perspectives. Châtelet's work, however, seeks to show, through examples from the history of mathematics, how gestures and diagrams play a pivotal role in mathematical invention. He wants to show how the material mobility of the human body comes to produce formal mathematics. Diagrams are essential clues for him because they provide a trace of the moving hand, while also enabling – on the surface of the paper – the exploration and creation of new objects and dimensions. Châtelet thus brings together two hitherto distinct areas of research – on gestures and diagrams, respectively – in pursuing his non-representational, non-dualistic account of mathematical thinking. The diagram, argues Châtelet, is by its very nature never complete, and the gesture is never just the enactment of an intention. The two participate in each other's provisional ontology. In contrast to current work around gestures on the one hand and diagrams on the other, Châtelet insists that extracting one from the other is both awkward and possibly misleading. He argues that the gestural and the diagrammatic are pivotal sources of mathematical meaning, mutually presupposing each other and sharing a similar mobility and potentiality.

To exemplify this claim, Châtelet selects certain episodes in the history of mathematics and physics to show how particular diagrams – what he terms 'cutting-out gestures' – have been deployed during inventive thought experiments to bring forth new mathematical concepts. Châtelet is careful, however, to analyse the way the mathematical concepts are folded into the material activity during the event, rather than simply ascribing the invention to cognitive ability or human discernment. In other words, he uses these historical episodes to explore ontological questions about the relationship between the mathematical and the physical, as well as cultural questions about what it means to do mathematics. He argues that the study of such diagram-gestures helps us undo some of the troubling consequences of the Aristotelian division between movable matter and immovable mathematics. We look at his examples – Archimedes, Oresme and Cauchy – and present his theory of mathematical inventiveness. In the last section of the chapter, we discuss a teaching experiment in which students watched a Nicolet stop-action film of a mobile circle, and then were prompted to diagram what they saw. We show examples of their diagrams and discuss how one might interpret their work in terms of Châtelet's ideas on the coupling of diagram and gesture. We use this experiment as a way to exemplify Châtelet's approach in the context of more modest mathematical breakthroughs and also as a way to illustrate how matter is always at play in the relationship between mathematics and the learner.

In Chapter 4, we examine some classroom episodes through this Châtelet-inspired materialist approach. In particular, we want to see how mathematical inventiveness in the classroom can be described not only in terms of virtuality, but also in terms of an assemblage of materialities (including children, teachers, computers, projectors, hands, arms, etc.) that operates more as a system of 'impersonal' forces and less as a reflection of individuated human agency. The two classroom examples discussed in this chapter are not meant to be prescriptions for classroom teaching and learning; instead, they are meant, once again, to underscore the entanglement of gestures, diagrams and words, and to put this concept of the virtual that we have adopted from Châtelet to work. However, within this pedagogical context, the diagrams involved will be the dynamic ones of the computer screen, which not only put mathematical objects into motion, but also give rise to new gestures and words. We are interested in how particular virtualities that might otherwise remain dormant are actualized through the use of the technology of the computer screen. More importantly, our materialist approach to these examples aims to focus on the role of movement and affect in mathematical activity and to resist a reading of mathematical behaviour that relies too centrally on language use.

Chapter 5 seeks to pursue our dedication to materiality by probing and stretching our readings of language use in the mathematics classroom. While Châtelet's philosophy of mathematics draws attention to the crucial role of gestures and diagrams in mathematics, offering powerful insights into the materiality of mathematics itself, he does not attend to the role of language in mathematics explicitly. This absence is problematic, because language plays a significant role in how the body comes to matter as a sociocultural entity, and it plays an equally important role in learning mathematics. While the study of materiality in teaching and learning mathematics helps us resist the logocentric or language-centric reading of mathematical behaviour found in many contemporary discursive approaches, one clearly still needs to attend to language use in studying mathematics teaching and learning. The challenge is to use semiotic and linguistic tools to study activity without blinding the researcher to that which might not lend itself to such an analysis. In Chapter 5, we experiment with ways of doing so, studying language use as part of a collective process of material *in(ter)vention*, not a translation of thought or only (or mostly) a semiotic act of representation. We propose this definition of language use in terms of in(ter)vention – rather than communication – to draw attention to both the materiality and inventiveness of language. In offering this material reading of language production, we hope to show how a *micropolitics* of classroom discourse attends carefully to the production of meaning while continuing, as in Chapter 4, to decentre the rational thinking subject as the source of meaning. This chapter pays tribute to the extensive and enlightening work that has been done by scholars of discourse analysis, such as Michael Halliday and Norman Fairclough, and shows how our materialist approach is not necessarily at odds with some of their basic orientations. Then, again drawing on Barad, but also on other contemporary philosophers, we dwell on the materiality of speech – the 'auditory gestures', as Rotman calls them – as a way of motivating the move from language as communication to language as in(ter)vention. We experiment with an alternative research methodology by which meanings in the mathematics classroom emerge in the pauses, accelerations, fallings-away and other bodily encounters that produce sounds, rather than merely in the discrete sonic units of spoken words. By contrasting this analysis with a more traditional discursive one, we hope to draw attention to some of the traps into which current transcription practices lead mathematics education researchers, as well as some of the openings for richer accounts of the motions and emotions of classroom activity.

In Chapter 6, we look at how the correlation of the 'true' and the 'sensible' (which are coupled in Rancière's political aesthetic approach) has served to produce Western conceptions of mathematical ability. Current theories of embodiment privilege the body in a way that has never before happened in the history of mathematics. If Western culture historically invested in a disembodied vision of mathematics, how will the stakes change if we embrace a more embodied one? In particular, what are the political ramifications of taking embodiment so seriously? We argue that the senses have played a unique role in conceptualizing mathematical ability, and we show how images of embodied (dis)ability operate in our educational practices. We tell one particular historical story, tracing how the concept of intuition is tied to particular Western assumptions about perception. Our aim is to show how post-humanist approaches to the body demand that we interrogate this notion of intuition, as well as the assumption that a body possesses a set of sense organs of pre-given capacity. Following a Deleuzian approach to the body as assemblage, the sensory organs are but one configuration of an unstable collective that may also include walking sticks, rings, cylinders and moving points. Sense organs are provisional and open to new configurations. We survey some of the increasingly critical literature on disability, which disputes disability as an assignation to the individual of a physical deficit.

We are sensitive to the ways in which this work on (dis)ability is polemical, but we believe that the arguments are worth studying for how they teach us about the body and how they make visible taken-for-granted exclusionary practices within education. We see in this work an attempt to move away from Kantian conceptions of perception-as-synthesis towards a study of the relational, highly variable and responsive rhythmic vibrations that are the foundation of sensation. In this sixth chapter, we explore the proposal that the body is thus constituted through thresholds or levels of resonance, and sensation is dislocated and only provisionally situated in a perception. Indeed, this is a body with potentially different perceptual capabilities from those which are currently considered normal. Hearing, touching, tasting, seeing, smelling and any other modality are temporary – if persistent – kinds of perception, but our future may entail entirely different calibrations of sensation. This post-humanist approach to sensation allows the body to break free from the confines of current perceptual organisation and demands that we recognize the human body in all its potentiality, even in our current classrooms, where bodies can be seen as differently abled and differently (organ)ised rather than disabled or distracted. This is not to dismiss the reality of those with disabilities, but rather to help researchers and educators think differently about the processes by which (dis)ability comes to be recognized in classrooms. We open the chapter with an example of a young girl interacting with a touchscreen application that shows how we can understand her various visual, auditory and haptic sensations as part of a rhythmic play beneath her ability to form concept-producing judgements.

Chapter 7 seeks to contribute to the political discussions around mathematics education more broadly. We recognize that any theory of embodiment must adequately address the sociocultural structuring of experience. The challenge is to attend to the materiality of experience, within the phenomenological tradition and contemporary post-humanist materialisms, while also recognizing the macropolitical and economic forces that in part shape it. Many political discussions within education hitherto, for the most part, have been informed and inspired by the critical discourse-focused approach of Michel Foucault. They inherit from Foucault an unwillingness to consider phenomenological approaches to the body – like that found in the work of Maurice Merleau-Ponty – because these seemed to essentialise and universalise the body and 'lived experience'. However, through the work of Jacques Rancière, who shows how one might effectively study the coupling of the phenomenological, aesthetic and political aspects of cultural practices like mathematics education, linked as this is with the later Foucault's work on dispositif, we investigate the way in which political participation and aesthetic practices are reciprocally implicated, not simply in terms of *judgements* of taste, but additionally in terms of the material distribution of what is taken to be *sensible* in mathematics. Rancière enables us to rethink the mathematical aesthetic, and its ensuing political implications, in terms of the practices of consensus that determine what is taken to 'make sense' in mathematics education. In this chapter we pursue acts of *dissensus*, both within mathematics itself and in relation to existing policies and standards for mathematics teaching and learning. In Chapter 7, we explain how this concept of dissensus allows us to bring together a discourse-focused approach with a new materialist treatment of the body. We draw on Tom Popkewitz's writings to highlight some current forms of consensus, which unexpectedly become more invasive through certain reform initiatives, and then we outline possibilities for dissensus. Although our focus is on the macroscopic, we wanted to show how political-aesthetic forces operate in everyday classroom activity, so we open the chapter by showing how consensus and dissensus are at play in an excerpt from a well-known video of a grade two classroom.

Chapter 8 returns to Châtelet's formidable notion of the *virtual* to argue that one can speak not only of the role of the body *in* mathematics thinking, but also of the body *of* mathematics. Through the powerful notion of the virtual, Châtelet shows that mathematics can and does straddle the supposedly independent worlds of the mathematical and the physical. In so doing, he challenges our conceptions of what a mathematical *concept* might be, which in turn leads to interesting and provocative ideas about what might be involved in *learning* concepts like number, and what repercussions this might have for the shape of the school mathematics curriculum. We focus in this chapter on how one might look at mathematics curricular concepts for how they are actualizations of the virtuality of the material world. Our aim in this chapter is to show how Châtelet's philosophy of mathematics offers a new way of animating mathematical concepts. Indeed, a theory of embodiment must allow for a mathematics that is alive and changing, rather than just a human body that is alive and changing.

In the Conclusion, we provide an overview of the central constructs used in this book, such as virtuality, assemblage, pedagogy of the concept, boundary-making practices, inventiveness and dissensus. We describe more concisely the main contributions of the book, particularly in relation to existing work in mathematics education. We also pick up some loose threads, with a particular focus on the important role that affect plays in the work of many of the theorists on whom we draw, such as Deleuze, Manning, Massumi and Sheets-Johnstone. We also recap some of the pedagogical possibilities that emerged from our work on both inventiveness and aesthetics in the classroom. These possibilities are framed less as suggestions for institutional change than as provocations for localized experiments in practice.

Synopsis

This book is first and foremost about a rethinking of school mathematics in terms of an inclusive materialist philosophy of mathematics. While such a rethinking has implications for teaching and learning, our book does not provide a specific theory of learning or a basis upon which to design textbooks and curricula. Neither does it resolve the many dilemmas involved in teacher education, assessment and the hegemonic role of mathematics in our society. That said, we hope that researchers, curriculum designers, teacher educators and critical theorists will find our basic philosophical stance, and our subsequent analysis of examples of mathematics teaching and learning, insightful and provocative in opening up new directions for research and practice in mathematics education. In particular, the book is meant to show how inclusive materialism might alter the way we think about the embodiment of mathematical concepts, offering alternate ways of studying how students learn concepts and of how we might choose and order concepts as part of a curriculum sequence. When concepts are animated differently, learning is similarly altered. Inventive acts in classrooms become part of a growing material assemblage, a process of embodiment in which the *potentiality* of the body is emphasized. In wanting to attend to the collective nature of these acts, and to the ways that non-human materials factor into the process, we were faced with challenging methodological choices. Embedded throughout the book are attempts to look at data differently and to reconsider what constitutes research data more generally. This kind of experimental work is important because of the way it forces us as researchers to reckon with the radical limitations of our research methods. Finally, the book considers political questions about embodiment and the regulative practices of school mathematics, and it points to new methods of analysis that might better attend to the ways (dis)ability is lived within school mathematics.

Although Chapters 1 and 2 describe the philosophical framework for the book, the first focusing on the body and the second on materiality, and subsequent chapters focus on different applications of this framework, readers can either read the book from beginning to end or read the chapters as stand-alone explorations of different issues in mathematics education, depending on their familiarity with the theory at hand and on their interests. We decided to include diverse examples in each chapter, from early childhood encounters with mathematics to advanced or expert encounters, with the aim of breaking down the division that often keeps these two kinds of activity apart. In Chapters 3 and 4, we unpack the important notion of the virtual as exemplified through inventive moments in classrooms. In Chapter 8, we return to virtuality, which is central to understanding our materialization of mathematical concepts, while discussing examples of advanced mathematics. Those who have a keener interest in sociocultural or discursive approaches in mathematics education research will find this considered in Chapters 5, 6 and 7, and also in Chapters 1 and 2 where we develop our theory of inclusive materialism.

1 When does a body become a body?

One common way in which embodiment is taken up in mathematics education is to focus on how students interact with and experience mathematical concepts through their bodies. For example, students may be invited to use motion detectors to move in such a way as to create a particular kind of distance versus time graph, thereby experiencing, through their own moving limbs, what slow, fast and stopped motion feels like on a Cartesian coordinate system. Similarly, students might be invited to use their hands and arms to mimic and/or create certain functions, sometimes also using their voices to signal crossings of the x-axis or relative maxima and minima. The involvement of the body can also be subtler, as in the dragging of the mouse across the screen, which requires both movement of the hand and of the eyes, in order to produce objects that vary and relations that remain invariant. Thus, the eyes, the ears, the arms, the legs and even the torso are evidently involved in the production, expression and communication of mathematical concepts. While some of these activities aim to emphasize the role that the body might play in developing mathematical understanding, some of them also suggest that such full-bodied experiences can motivate students to engage in mathematical activity and challenge their beliefs that mathematics is an abstract, disembodied and depersonalised discipline.

This book aims to unpack the assumptions and explore the consequences of this kind of research. We ask, first and foremost, how the physical aspects of mathematical activity – be it that of students or mathematicians – are transformed into the so-called abstractions and generalisations of formal mathematics. Are these bodily activities somehow internalized first, and then only later subject to particular mathematical or classroom discourses? Is it just a matter of feeling what a slope is like and then matching this feeling to certain culturally available signs? Or might the materiality of the

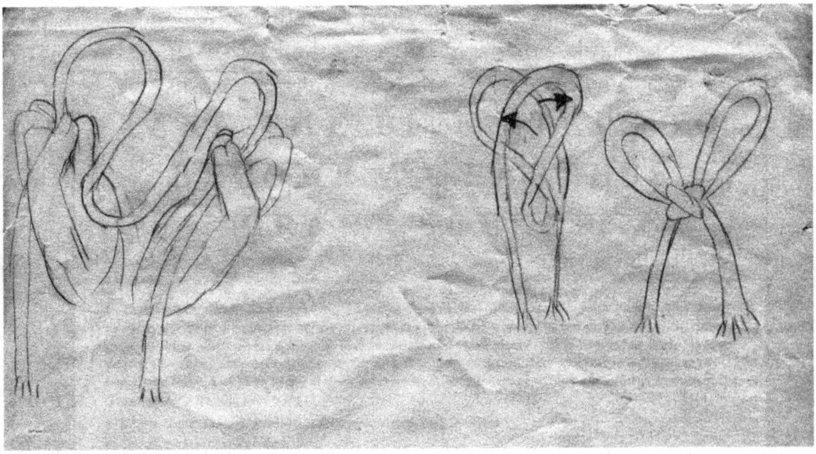

Figure 1.1. Student drawing of knots.

actions and the mathematical concepts be mutually entailed or entangled in ways that move beyond a theory of representation? Consider the case of a grade six student who is asked to create a diagram of the knot she has produced using string. As can be seen in Figure 1.1, she draws *not* just the knot, but also her own hands, which convey the manual process by which the knot came to be.

The hands in this image also orient and position the viewer so that the body of the viewer is hailed and addressed. In what ways do more sophisticated mathematical diagrams continue to hail the embodied viewer? And might this form of address actually entail a material coupling with the human body so that the entire assemblage of diagram, string and sense organ is bound up with the mathematical concepts under investigation? Rather than thinking of the diagram as a representation, what happens when we imagine it as generative of a concept? Does mathematical activity actually entail a remixing of matter and meaning in such a way as to reconfigure the world? Indeed, might such a vision of mathematical activity allow us to better study its creative or inventive aspects? In this chapter, we focus on how such questions cause us to rethink the contours of the mathematical body. Rather than study the learner who drags the triangle on the screen as an enclosed body that knows, acts or feels independently from the mouse, the screen or the digital interface more generally, we propose a theoretical framework that looks for how bodies are assembled through activity. Roth (2011) offers a simple but powerful example of embodied knowledge when he describes the common experience of not being able to remember a phone number of a long lost friend but, upon being given an actual telephone, finding that one's fingers remember the right dance on the keys. In the same way that we can think of the finger, the hand and the eyes as knowing, might we also be able to conceive of bodily encounters whereby 'knowing' extends beyond the boundary of the skin? Might it make more sense to think of the body as a growing and contracting assemblage of diverse materials?

In Chapter 2, we begin to examine what might constitute these diverse materials, but here we focus on the question as to when a body becomes a body, a question that will remain open throughout this chapter. We pose this question so as to trouble assumptions not just about the nature of the body, but also about the very determination of bodies as individual entities. Asking 'when does a body become a body?' directs our attention to the processes by which bodies – be they human or non-human – come to be counted as bodies, processes whereby a body is recognized as a body. In the context of mathematics education, increasing interest in the role of the body in teaching and learning has made such reflection timely and important. Focusing on the ongoing processes of embodiment allows us to track diverse forces at work in mathematical activity, at various scales, and to show how those forces co-mingle in complex ways, suggesting that a body is less an entity and more a process of becoming. Our aim is thus to de-essentialise the body as traditionally conceived, perhaps even to fragment and decompose it so that it becomes, if only momentarily, unrecognizable or at least unfamiliar, allowing us to rethink how bodies are assembled and incorporated into the world. In this chapter, we explore different theories of embodiment, comparing them with and contrasting them to the approach we wish to pursue in this book. These diverse theories include: neurobiology, feminist philosophy, Foucauldian theory, discourse analysis, materialist phenomenology and the notion of the extended self. We cannot possibly do justice to all the nuances of these positions, but our aim is to begin to map the theoretical terrain and focus on the extent to which these theories move away from an essentialised view of the body.

Theories of embodiment

Mathematical thinking and learning involves the use of the body – the brain, after all, is part of the human body. Over the past two decades, though, a variety of scholars have been pursuing the question of what the non-brain parts of the body might have to do with mathematical thinking. This pursuit has taken on many different forms, each of which makes different assumptions about what the human body is and how it interacts with the material and social contingencies of mathematical experiences. Lakoff and Núñez (2000), for instance, working within an embodied cognition paradigm, focus on how conceptual knowledge is mapped onto the sensorimotor system, claiming that the semantic content of mathematical concepts can be understood in terms of the way human bodies function in the world. Some scholars criticise such an approach for how it downplays the role of the social, while examining only those mappings from the non-brain parts of the body to the brain. Working within mathematics education, Radford (2009) has further developed this approach by proposing 'sensuous cognition' as a construct for studying a more multimodal conception of learning – one that is not confined to the metaphoric mappings of Lakoff and Núñez, but also invokes the semiotic coordination of speech, body, gestures, symbols and tools. Here, the social and material worlds feature strongly as components of embodiment, and specific sensorimotor actions, such as gestures, are studied not just as initial sources of mappings, but also as ongoing means of understanding and communicating. The qualifier 'sensuous' seems to encompass that of 'embodied', while continuing to treat the human body and the material world as interrelated but distinct elements of a system.

Theories of material phenomenology (Hwang & Roth, 2011) have begun to break down some of these distinctions, arguing, for example, that certain phenomena that are usually described as 'mental' (in the brain) might be better thought of as 'corporeal'. Such an approach dissolves the dichotomizing, metaphor-based theory of Lakoff and Núñez, offering the word 'flesh' as an alternative to the dualistic tradition of the mind/body split. Complexity theorists and enactivists (e.g., Davis & Simmt 2003), drawing on systems theory (e.g., Maturana & Varela, 1987), have perhaps gone furthest in dissolving fixed concepts of the human body, defining it as an ecological system sustained through boundary negotiations. The human body might thus be more or less than its visible, physical contour, depending on the situation. These scholars have revitalized the human body as an active force in learning, showing how human thinking involves various parts of the body rather than only the Cartesian 'mind'. Their theories encourage researchers to attend to new phenomena in the mathematics classroom, including the multimodal ways in which teachers and students communicate with each other.[1]

Stevens (2012) suggests that the terms *embodied* and *embodiment* need to be foregrounded and unravelled, seeing as they have been taken up by a

[1] See, for instance, the 2012 special issue of the *Journal of the Learning Sciences, 21*(2).

diverse set of authors all with the aim of addressing the role of the body in teaching and learning mathematics. He asks what it means to say that mathematical cognition is embodied. Indeed, the phrase 'cognition is embodied' is criticised by Sheets-Johnstone (2009) because of the way it demotes the body to acting merely as the vessel or container of some higher act of cognition. Stevens (2012) proposes that one can distinguish two dominant approaches in the field – conceptualist and interactionist – and that most of the work on embodiment tends to take up these two approaches to a greater or lesser extent. The first approach is strongly linked to cognitive linguistics, in which it is assumed that individual human beings share a common biology and 'thereby develop common internal concepts and conceptual systems based on these experiences' (p. 338). According to this approach, concepts are abstractions of common phenomena, which lead to 'phenomenological primitives' and eventually to an individual's conceptual system (p. 338). For instance, Alibali and Nathan (2012) claim that mathematical cognition 'is based in perception and action, and it is grounded in the physical environment' (p. 248). Aligned with the 'conceptualist' approach, these authors interpret the material actions of students and teachers as external simulations of some prior or primary internal conceptualization. Thus, their use of 'based in' and 'grounded in' seems to construe a binary relationship between mind and body, even as the authors claim to do otherwise, because mathematical concepts are assumed to dwell inside the mind, while the body bumps against the world in ways that trigger internal acts of synthesis.

We find that the 'conceptualist' approach to the body does not do justice to bodily activity because it ultimately demotes activity to simulation rather than full-bodied thinking. The second approach comes from the study of interaction and reflects a more social perspective on processes of meaning making. Within the interactionist approach, Stevens (2012) advocates the study of gesture as one among many 'modal resources for producing meaning and action within interaction', while also examining the way that gestures are often 'designed' for purposes of communication and pedagogy (p. 340). The work of Ricardo Nemirovsky falls into the interactionist tradition. It is inspired more explicitly by phenomenology, especially as it evolved in the work of Maurice Merleau-Ponty, where bodily activity is studied in its own right, rather than as a reflection of an interior mental concept, and where 'mathematical insights developed by learners are expressed in and constituted by their perceptuo-motor activities' (Nemirovsky, Rasmussen, Sweeney, & Wawro, 2012, p. 311). Nemirovsky et al. speak of perceptuo-motor possibilities immanent to the situation and focus on the human capacity to intend or intuit or imagine these possibilities with reference to a 'state of readiness to notice, move, react, expect, and so forth' (p. 292). This work shows us how micro-ethnographic facets of experience impact and partially shape the ways we teach and learn, but it tends to centre human will or intention in the orchestrating of experience. In contrast, our aim here is to focus less on human intention and more on distributed agency.[2] We want to problematize some of the ontological tenets underpinning particular conceptions of the human body as the principal administrator of its own participation. Moreover, we find in neither the conceptualist approach nor the interactionist approach, as characterized by Stevens, adequate treatment of the biopolitics of the body. Neglecting the political dimension of the body all too often leads to educational theory that simply cannot speak to the material conditions of schools.

The recent research on 'embodied simulation theory' (Gallese & Lakoff, 2005), for instance, is problematic in this respect. If we identify thinking as a 'bodily activity', we might be tempted to study the brain as the legislator of such bodily activity. Theories of embodiment often lead to neuroscientific accounts of learning because the brain figures prominently in human and animal activity. Such an approach aims to show how the sensorimotor system is generative of conceptual knowledge, arguing that conceptual knowledge is 'nested in the activity in the brain', where the activation of different neural cortical networks 'can enable the capacity to distinguish, recognize, categorise, and ultimately conceptualise objects, events, and the state of affairs in the world' (p. 3). We are enthusiastic about this research in its attempt to centre affect (preconscious perception) and a kind of knowing-in-the-body, but we are concerned that research on brain activity makes possible reductionist and deterministic claims about neurobiology as the fundamental architecture of learning.[3] Such an approach may devalue sociocultural factors involved in learning. Although valuable attempts to bridge the sociocultural with the neurophysiological have been made (e.g., Hall & Nemirovsky, 2011), and although there is much that can be learned from both, one can point to particular methodological assumptions in neuroscientific research that seem to undercut a sociocultural approach to the

[2] Here we build on the tradition of distributed cognition (Edwin Hutchins) that originated in the mid-1980s. Unlike developments in phenomenology that follow the early Merleau-Ponty, for whom embodiment was tied to human bodies, Roth (2011) argues that new trajectories in material phenomenology, in particular those found in the work of Michel Henry, adequately treat the body as a distributed network.

[3] Of course, neurobiological claims are not *necessarily* reductionist and deterministic, as they need not be intrinsically opposed to social, cultural and kinaesthetic aspects of human experience. We are merely pointing to the tendencies of the current research culture in the sciences.

body. The emphasis on the individual, for instance, rather than the mutual construction of intersubjectivity, and the focus on pathology, rather than creative collaborative interaction, threaten to displace theories of psycho-social development by means of neuroscientific theories of visual/motor development. And many of the claims (and desires) of this research are not yet adequately supported by clinical data. We are concerned, for instance, by arguments based on mirror neuron research that claim perception and action are coupled in ways that suggest a capacity for empathy is biologically determined or that we have a direct experiential grasp of the minds of others through neuronal mirroring (Gallese & Lakoff, 2005). We find Sheets-Johnstone's (2012) critique of claims based on this research very important in that she underscores the fact that 'mirror neurons are contingent on morphology and corporeal-kinetic tactile-kinesthetic experience' and that the entire activity of mirror neurons is actually conditional on kinaesthetic activity and sociocultural factors (p. 1).

The issue of how to make sense of findings from neurobiological research has been taken up more generally in feminist philosophy, where the tension between biology and political theories of subjectivity has always been a source of concern and inspiration. Feminist philosophy thus provides fruitful ground for a questioning of essentialised views of the body. This tradition fought hard to show how the body was a product of socio-cultural forces and insisted on studying identity as performative (what we *do*), rather than essentialised (what we *are*). These approaches to the body and its relation to social phenomena pushed the discussion of embodiment beyond the reductive positivism of behaviourist research from previous decades. Judith Butler's *Bodies that Matter* in 1993, Elizabeth Grosz's *Volatile Bodies* in 1994, as well as numerous 'body' readers argued that the body be recognized as a non-essentialised, discursively construed performative entity rather than a biological given. Butler's work on subjectivity and the body problematizes the biology/culture divide. She shows how *both* sex and gender are designations that reflect cultural practices whereby the body comes to matter in particular ways. Drawing in part on Lacanian ideas about identification and the mirror stage, Butler focuses on the disconnect between a body-image and a phantom 'real' body, highlighting how the borders of the body-image are indeterminate and unstable.[4] Grosz

[4] For early Butler, processes of identity formation – who am I? – occur at the mirror stage when the 'real' body also 'comes into being' through a fundamental self-division and mis-recognition of the body-image. As the following quotation makes clear, Butler's account of the body is highly relational, both within the 'real' body and with other bodies: 'The mirror stage is not a *developmental* account of how the idea of one's own body comes

(1999) defined the body in similar terms, although she later turned away from this Lacanian framing. Of interest in Grosz's (1999) description of the body is an emphasis on processes of cohesiveness whereby a body is given unity. She suggests that the 'concrete, material, animate organization of flesh, organs, nerves, muscles and skeletal structure' contains the incomplete 'raw materials' on which a 'psychical and social inscription' is ordered and administered (p. 243). The body, according to Grosz, becomes a human body when it 'coincides with the "shape" and space of a psyche, a body whose epidermic surface bounds a psychical unity, a body which thereby defines the limits of experience and subjectivity' (p. 244). Thus, the human body is a non-essentialised body that has become inscribed by a set of socially coded meanings: '[I]t is indeterminate, amorphous, a series of uncoordinated potentialities which require social triggering, ordering, and long-term "administration," regulated in each culture and epoch by what Foucault has called "the micro-technologies of power"' (p. 244).

We can see in this particular line of feminist writing a strong focus on the power of discourse to delineate and mark the borders and surfaces of the body. We can also see how Foucault's work on power and discourse informs this approach to the body. Foucault studied power as a productive force, one no longer only associated with sovereign power or central authority but, instead, with discursive behaviour and its capacity to produce diverse material effects alongside the structuring hierarchies of state control (Krause & Rölli, 2008). This work is relevant to our reading of theories of embodied cognition in mathematics education, which often neglect the social constructs that in part structure all educational experiences, be they gender, class, race, sexual orientation or more broadly conceived social constructs, such as subjectivity and power. Núñez, Edwards and Matos (1998) make this very point, arguing that situated cognition is not about situating 'mind-free truths in meaningful contexts', but rather about 'seeing how the human creation of mathematics arises from human sense-making that is not arbitrary because it is bodily grounded [...] This means that in order to really make sense of mind-based mathematical ideas as fully embodied and contextual, one needs also to provide an understanding of the historical

into being. It does suggest, however, that the capacity to project a *morphe*, a shape, onto a surface is part of the psychic (and phantasmatic) elaboration, centring, and containment of one's own bodily contours. This process of psychic projection or elaboration implies as well that the sense of one's own body is not (only) achieved through differentiating from another (the maternal body), but that any sense of bodily contour, as projected, is articulated through a necessary self-division and self-estrangement' (Butler, 1993, p. 71, *emphasis in original*).

processes through which embodied ideas emerged' (pp. 23–24). We suggest that their call has yet to be adequately addressed in the embodiment literature in mathematics education, in part due to the difficulty of addressing questions about the body at multiple scales. How does one talk about the 'perceptuo-motor possibilities' of the body while also addressing the social entanglement of bodies? Approaches that focus exclusively on the individual learner or their physiology fail to contend with the latter. And yet, approaches that focus on the sociocultural are often criticised for not attending to the particular micro-practices of mathematical activity. To some extent, this is a matter of the time constraints on scholarship, because doing justice to either demands a sustained treatment. And yet, as researchers we would be remiss to simply abandon all attempts to study the ways in which these different spatial and temporal scales are interwoven. In this book, we try to show how work in these two areas might fruitfully be brought together, and we offer a new theory of the body that might better address the way meaning and matter are entangled.

Material phenomenology and post-humanism

We are particularly sympathetic to sociocultural theoretical approaches in mathematics education, but we find these are often hobbled by conceptions of discourse that limit all ontological claims about the body. Sfard (2008), for instance, redefines thinking as an embodied activity and effectively shifts the focus from individual cognition to social interaction. But her analysis of mathematical activity remains almost entirely language-centric. The strength of Sfard's approach – if we take it to be a turn to discourse and communication – also points to a weakness, namely its inadequate theorizing of the body. Hwang and Roth (2011) identify this issue in sociocultural studies more generally, arguing that 'logocentric' interpretations of Vygotsky's approach have often conflated or reduced thought to language, and thereby have eclipsed the body and its role in learning (p. 25). The new materialisms we discuss in Chapter 2 emerge in part as a response to this tension in the sociocultural approach:

> For new materialists, no adequate political theory can ignore the importance of bodies in situating empirical actors within a material environment of nature, other bodies, and the socioeconomic structures that dictate where and how they find sustenance, satisfy their desires, or obtain the resources necessary for participating in political life. This is in fact something that feminists and class theorists have often insisted upon, and we would add in

> this context only our concern that such material dimensions have recently been marginalized by fashionable constructivist approaches and identity politics. (Coole & Frost, 2010, p. 19)

Much of the research on mathematics student identity, for instance, is powerful and compelling precisely because of the way it shows how individuals and institutions produce or enact identity *through discourse*, and yet such an approach risks abandoning the body if it treats physical matter as entirely inert. (We explore this issue in the next chapter.) In our estimation, the question has become how the body might figure in a more active way within these discursive approaches to teaching and learning.

The European phenomenological tradition, which includes the work of Edmund Husserl, Martin Heidegger, Maurice Merleau-Ponty, Jean-Paul Sartre, and other contemporary philosophers, has emerged as a powerful theoretical framework for studying embodiment and mathematics. Phenomenology is distinguished from other philosophical domains by its study of the structure of lived experience, evolving as a field of inquiry in the twentieth century, primarily through the work of Merleau-Ponty, to focus on human perception, sensation and affect. One of the concerns with this tradition has been its over-emphasis on human intentionality as the determining factor in experience. In the context of mathematics education, Wolf-Michael Roth develops and applies his own version of material phenomenology in an attempt to address the role of the body in learning. For instance, Roth (2010) studies a young student's tactile and multimodal engagement with a cube, where he argues that the movement of the child's hands erupts or emerges without intention or governing concept. According to Roth, these haptic encounters are more originary than language, somehow detached or free from an intention of will. It is *in* the hand that the memory of prior encounters with cubes is immanent. Roth suggests that a pre-verbal 'I can' coordinates this encounter with the cube, and that the social world itself begins to emerge through touch and the coordination of movements between eyes and hands.

We can see in this analysis how Roth centres the hands in the doing of mathematics and decentres the child's will or intention, ascribing activity instead to a pre-verbal 'I can' that shifts our attention to the capacity of the body to effect change. This analysis focuses centrally on the perceptible, touchable cube, studying the student's encounter with what is commonly taken to be material and sensible, namely the physical manipulative. While Roth's cube example sheds light on the role of the body in learning, the analysis fails to do justice to the materiality of either the cube or the mathematics; that is to say, it fails to reckon with the way in which the cube is itself becoming-cube through its encounter with the child, shifting its own boundaries in this process of becoming. Roth treats the non-human material in this encounter as passive and inert without adequately attending to the potential pre-verbal 'I can' that might emanate from other sources. Moreover, the mathematical concept of cube remains untouched and untroubled by the encounter, as though it were indeed an immaterial and inflexible concept that happens to be somehow manifest in this particular instance.

In the description of the child's activity, Roth speaks of the 'body' in terms of the isolated human body, while the cube is that which is acted upon. In this book, we ask: What happens to our understanding of learning when we consider the cube itself as not merely passive, but actively involved in the assembling of meaning? How might we conceptualize a material agent or recognize the way that degrees of agency saturate the situation and all its 'actants'? Instead of associating agency with human will and intention, or assigning an 'I can' to only one agent, we want to broaden the concept of agency so that it includes the force of affect and the potential energy of relations. Only by doing so can we begin to think about the ways that the human and the non-human form bonds that reconfigure the world. In a similar fashion, one might consider a man walking in a dark room with a stick:[5] Does the man feel his hand touching the stick or does the man feel the end of the stick touching the contours of the dark room? Distributing the agency across the cube situation demands that we rethink the borders of the body, the nature of matter, and the ontology of mathematics. What if we consider the assemblage of child-cube-concept to be *a body* that emerges at that instant? Might such an approach allow us to study the event of learning in new ways?

Following Harroway (2008) and other post-humanist philosophers, we suggest that the human body is not fixed or finished and, more importantly, that the human body is not the only agent involved in processes of learning. The matter of the cube and the matter of the mathematical concepts are also agents in this context, as are the teachers and their policy-inflected pedagogical actions. This is a philosophical claim about the nature of embodiment and the nature of learning, as well as the ontology of mathematical concepts. We focus more closely on the last of these claims in Chapters 2 and 8, while in this chapter we want to stretch this phenomenological analysis so far that it breaks with its centring of the human agent.

[5] Neils Bohr uses this example, as well, as have many others, including Bateson (1972).

Our hope is that in doing so we can begin to address the diverse forces that are at work in any experience of learning and move beyond what many consider to be the overly narrow focus of phenomenological studies. We propose that the body is an assemblage of human and non-human components, always in a process of becoming that belies any centralizing control. The body in/of mathematics partakes in a 'relational ontology' or 'mutual entailment' that binds the components together in a process of becoming embodied (Barad, 2003, p. 820). In the example discussed earlier in this chapter, the body that is becoming is an indeterminate assemblage of child, cube and concept. Doubtless, there are other threads entangled in this becoming, but we are using this very limited scene to rethink the body outside of traditional models.

Malafouris (2008) offers a useful conceptual bridge between the biological and the cultural, and he does so by reframing the notion of the body in terms of material engagement. In his concept of the 'extended self', Malafouris envisions the boundary of the taken-for-granted-I as being 'changeable and extendable to the outside world rather than fixed at the surface of the skin' (p. 1994). This new concept of the self (and of the body) eschews the notion of an isolated human agent acting upon the inert and inanimate world. By extending the self, Malafouris invites us to move beyond the 'container' metaphor of the body and consider the 'self that is located neither inside nor outside the brain/body, but is instead constantly enacted in-between brains, bodies and things and thus irreducible to any of these three elements taken in isolation' (p. 1997). Instead of focusing on the embodiment of cognition, which does not 'succeed in dissolving the ontological bounds of the *res cogitans*' and which 'remains, more often than not, trapped inside the biological bounds of the individual' (p. 1997), Malafouris wants to look at how the self is 'enabled through the act of embodying':

> More simply, if there is such a thing as the embodied self, then it is a self that constantly projects and extends itself beyond the skin actively engaging and incorporating its material surroundings via the interface of the body. Embodiment is not a delimiting property – far from it. It is instead the main perturbatory channel through which the world *touches us*, is *attached to us* and even *becomes part of us*. (p. 1997, *emphasis in original*)

Malafouris offers both archaeological and neuroscientific evidence of the effects of the incorporation of inanimate objects into the body 'schema' – objects such as rings, swords and tools. Not only does our use of inanimate objects provoke remapping in the brain, but the ordinary personal possession of these inanimate objects can become 'coextensive with our body' (p. 1998). In pursuing the ramifications of acts of embodying, Malafouris insists on 'a continuous and interactive coordination between neural and extra-neural physical resources' (p. 1999). When one thinks of learners calculating on an abacus or drawing a circle with a compass, the implications of this line of thinking for mathematics education research become obvious: These tools become parts of the learner, continually changing the very constitution of their bodies. Going just a little further, for example, the common practice of banning the calculator on a test becomes not so different from covering the eyes or tying up the arm. Human bodies are constantly encountering, engaging and indeed amalgamating with other objects; the limits of our body are extended through these encounters (de Freitas & Sinclair, 2012).

While Malafouris argues for extending the self, or pushing out body boundaries, one might also consider how the body may contract. For example, Sacks (1984) describes how, having suffered a severe knee injury, his own leg became an alien thing to him, not seeming to belong to his own body. This contraction of the body has also been described by breast cancer patients, for example, who talk about their breast as not being a part of their body. Of course, the question of where the body begins and ends is also at play in determining the status of the foetus of a pregnant woman.[6] While examples of extension might be more prevalent in studies of material culture, contraction is quite central to disability research, which often focuses on bodily deficit. In Chapter 6, we argue for a post-humanist approach to the body that problematizes the pre-given capacities of organs, limbs or neurological connections, and thus sees contraction less in terms of what might be missing in a particular individual body and more in terms of different calibrations of sensation.

Objects, quasi-objects and agency

We begin this section with an example that sets the stage for how one might rethink the body and agency in terms of heterogeneous assemblages and processes of becoming. We draw on Radford and Roth's (2011) analysis of the 'space of joint activity' that students create to solve mathematical problems, as we find that it resonates in many ways with our approach, while also highlighting important differences. The authors describe how three students create their joint space 'through a complex array of corporeal

[6] We thank Ricardo Nemirovsky for drawing our attention to this important dimension of the non-essentialised body and to the contexts for which contraction might be as relevant as extension.

dimensions around the artifacts: this array includes a sophisticated coordination of very different sign forms (i.e., gestures, words, actions, and perceptions)'. Thinking is thus said to appear 'as something *collective*, as something *inter-corporeal* and *inter-objective*' (p. 233, *emphasis in original*). Their example involves grade four children engaged in a modelling task based on this situation: For his birthday, Marc receives a piggy bank with one dollar. He saves two dollars each week. At the end of the first week he has three dollars; at the end of the second week he has five dollars and so on. Using bingo chips of two colours (blue and red) and numbered plastic goblets, the students are asked to model the saving process until Week 5. Then, they are required to answer questions so as to find the amount of money saved at the end of Weeks 10, 15, and 25. Eventually the students develop a model of the situation in which the number of chips in the goblet at Week n is $1 + 2n$.

The authors describe the initial strategy of the children, which is to 'start placing the bingo chips in the goblets (three bingo chips in the goblet that correspond to the piggy bank of Week 1, etc.),' and the authors write that:

> Although interesting, the model proves to be of limited use to answer the questions about the amount of money saved in some distant week (like Week 25). Indeed, the bingo chips pile up inside the glass, making it hard to discern any structure, let alone a mathematical one. The students' attention is directed to the sequential additive actions (adding two bingo chips) that remain unsynthesised in a more abstract multiplicative structure. (p. 232)

Before rushing to the question of whether students successfully synthesized 'a more abstract multiplicative structure', we want to pay attention not simply to the material world involved in the activity, but also to the body of mathematics that is being shaped both by the children and the researchers. We want to focus on how the material world – the chips, the goblets, the limited surface of the desk – is implicated in this episode and how our reading of the episode also effects a boundary-drawing whereby the abstract and the concrete are delineated, as are the human and non-human. The chips and the goblets are seen by Radford and Roth as useful representations by the researchers, but we want to suggest that they operate as more than just representations that supplement the activity. We want to be able to talk of the students-and-chips-and-goblets-and-concepts as part of a growing indeterminate assemblage that remains incomplete in some fundamental way. After putting one blue chip in the goblets, the children begin placing red chips, two at a time. Although this enacts the process, it

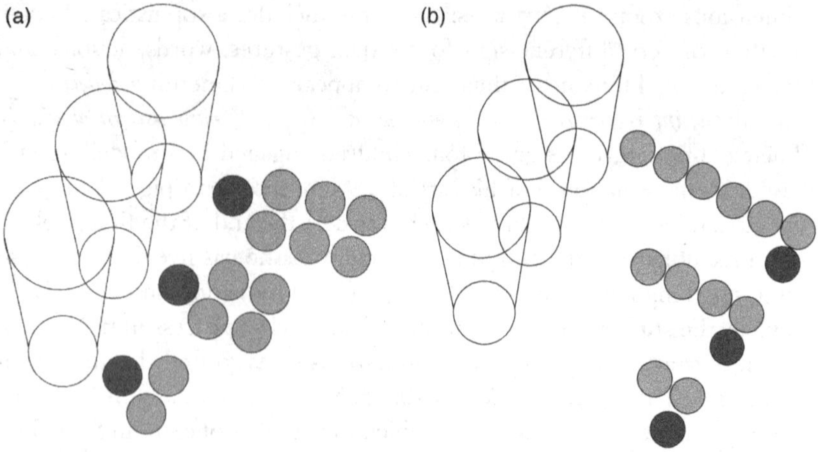

Figure 1.2. (a) Placing chips beside goblets; (b) another possible rearrangement.

does not result in a visualization that can be used in reasoning about the general case, and so the children are encouraged to put the chips in front of the goblets. Had the goblet instead been a flat and wide plate, the students would have been able to see (and count) the chips *on* the plate. By directing the students away from the containing possibility of the goblet, the students must interact with the goblets as placeholders, *beside* which they place chips. Further, the placing of the chips on the desk directs the material activity to the surface on which all the goblets rest, and thereby generates an absolute plane of reference, which then underscores the temporal break between subsequent weekly additions of two dollars. What one sees on the desk (Figure 1.2a) is not the dollars in the piggybank after a certain number of weeks, but the array of pairs of chips beside the piggybank acquired each week.

Eventually, the students run out of room on the desk and must extend the goblets and chips to the neighbouring desk. The surfaces on which the students work are thereby also part of the material practice; it enables the visual sequentialising of actions. What the students read as a temporal process eventually becomes a static, visual one in order to enable the discernment of static structure, but that discernment is entirely conditioned by the material assemblages formed during the activity – the grouping of the desks, the shape of the containers, the colours of the chips. As the authors describe, the students manage, through their physical work, to determine the amount of money saved on Week10, but do not, on their own, develop a mathematically optimal strategy. At this point, the teacher joins the activity and asks the students to describe their iterative, and therefore somewhat cumbersome, strategy. She asks the students what they notice about Week 5, pointing to the goblet, and the number of red chips, pointing to them. She repeats this question for Week 4. One student exclaims, 'It's always twice!', and the students discuss this doubling relationship by naming its particular instantiations ($1+1=2$, $2+2=4$, $3+3=6$, etc.). The teacher stays with the students, further working with them to transform this symbolic relationship into a mathematical expression relating the week number and the number of chips. She does this by grabbing the Week 5 goblet and saying 'five', and then pointing to the red chips and saying 'times?' in a questioning way. A student responds 'two' and the teacher points to the blue chips, asking 'plus?' The same student says 'one'. She then takes that same goblet and moves it to a place that would correspond to Week 10, and asks, 'What would you do for Week 10 if Week 10 was here?'

Radford and Roth describe the joint activity of the teacher and students as 'togethering', stressing that the students and the teacher have different goals for their activity, but also emphasizing that, within their joint activity, there is no dividing line between the realm of the students and that of the teacher. This approach enables the authors to account for the temporal continuity of culture and its transmission (in this case, the development of an algebraic structure) without reducing learning to acquisition. For Radford and Roth, the teacher and students are inseparable; they cannot be reduced to activities of teaching, on the one hand, and learning, on the other. From our point of view, this is obvious in the way that the teacher breaks through and redistributes the environment by coupling the weekly number positions with the number of chips, effectively forging a connection between two previously independent arrangements and extending the coupling to virtual positions on the table. She also incorporates the operations that will link the goblets and chips as she intones the 'times?' and the 'plus?' that nudge the students into a refrain of *times two plus one*, uttered to the beat of grabbing and pointing hands. If the activity is successful, it is in no small part because of this new assemblage.

The multiplicative structure that emerges is not so much a repeated addition ($2 + 2 + 2 + \ldots$), which could have arisen from the rhythmic placing or counting of pairs of chips, but comes to be a doubling of chips through a reconfiguring of the chips and a redirection of the gaze from cross-week to between-week-and-chips. The orthogonal flight of the eye by the pointing of the teacher's hand transforms the assemblage of students-and-chips-and-goblets-and-concepts-and-teacher. The concept of multiplication is captured in part from the rectangular configuration of two rows of n chips, where the multiplicity becomes the planar region occupied by the chips. But had the students placed the chips differently, as in the hypothetical Figure 1.2b, the assembling of matter and meaning would have changed. The number of chips accumulated each week would form a linear relationship, and this 'mathematical meaning' would be coupled with the particular material conditions, which would have entailed different teacher gestures and intonations. And yet, in either case, to explain how the students ultimately generate the desired algebraic expression for the number of chips in Week *n*, one still needs to explain how the actions of the students and teacher in their material world can produce the apparently abstract mathematical expression, which seems to have nothing to do with chips or dollars, desks or pointing fingers. Does mathematics really just stand there, silently waiting for the breakthrough in insight or shift in attention? Or might it somehow be much better implicated in the moving hands and the configuration of chips? If so, what do we mean when we say that the actions are concrete and the mathematical expression are abstract?

We are certainly not the first to try to think of the body in post-humanist terms. Over the last 25 years, Bruno Latour, who has worked extensively on the history and sociology of scientific practice, has argued a related point. Developing a post-humanist sociological method known as Actor Network Theory (ANT), Latour has railed against the modernist tendency to divide the world into two spheres with their different agencies: the human sphere (a sphere of originary freedom, will, intention, agency, diversity) and the sphere of nature, or the external world (a sphere of limited to no agency, acting with mechanical precision). It is precisely this division, argues Latour, which hobbles our work in the 'social sciences' by enforcing dualistic theories of mind and body. One can indeed trace this dualistic legacy in much of our educational theory, especially those theoretical frameworks that construe learning as a dialectic between cognitive schemas (making patterns, generalising) and material objects (cubes, chips, goblets). Instead, argues Latour, the distinction between sentient and non-sentient matter is a question of degree, or perhaps of intensity. We follow Latour in his desire to dissolve the ontological divide between human thought and that which is outside of thought, and in his attempts to reconsider agency as distributed across a heterogeneous assemblage of diverse agents, both human and non-human. Consider, he suggests, a world composed of 'actants' or 'quasi-objects' (like people, cubes, goblets, numbers) that all partake in some degree of agency, as though a network or assemblage of catalysts or mediators. An actant is a source of action: 'something that acts or to which activity is granted by others. It implies no special motivation of human individual actors, nor of humans in general' (Bennett, 2010, p. 9). An actant may appear 'sometimes as a thing, sometimes as a narrative, sometimes as a social bond, without ever being reduced to a mere being' (Latour, 1993, p. 89). Even inanimate 'objects' are seen as active mediators in a social-material network, something that we will put to work in later chapters when we look more closely at mathematical concepts. A quasi-object is 'as much force as entity, as much energy as matter, as much intensity as extension' (Bennett, 2010, p. 20). One can see in this approach a commitment to rethink the mind/matter split.

This is not to suggest that there are no ways of distinguishing between human actants and non-human actants. Indeed, such a distinction is crucial for Latour and others who have used this approach to look carefully at how power is marshalled by particular actants. Latour's detailed histories of science effectively document the way human actants wield power and effect material change. In his study of European scientists in the seventeenth century, for instance, Latour (1990) shows how diagramming practices actually constituted and controlled what was taken to be visible (and invisible) and how such practices were indeed part of capitalist, imperialist and commercial interests. He notes how social and physical scientists gain status when they mobilize their preferred inscriptions and gather the gaze of others to these inscriptions: 'Scientists start seeing something once they stop looking at nature and look exclusively and obsessively at prints and flat inscriptions' (p. 15). Thus, the human body operates with particular force in this flat ontology, but it does not rise above or outside of this realm of things.

Drawing extensively on Latour, Bennett (2010) argues for a 'vital materialism' that might pursue a new kind of political theory. One might conceive of this political theory as a form of environmentalism in its attempt to recognize the political voice of non-human agents. But it pushes beyond anthropocentric images of the environment and takes a more radical position in de-centring human agency. Both Latour and Bennett are developing a vision of politics that aims to follow the way that agency and power are distributed across various material assemblages. For instance, Bennett uses Spinoza's concept of 'conatus' to explore and argue for a new theory of embodiment, where conatus is the 'active impulsion' of a body, the tendency to persist or persevere in 'its own being' (Bennett, 2010, p. 2).[7]

[7] For more complex assemblages, conatus is the 'effort required to maintain the specific relation of "movement and rest" that obtains between its parts, a relation that defines the mode as what it is' (Bennett, 2010, p. 22). In this sense, conatus refers to the dynamic entwinement of force and matter, quite literally the relation of movement and rest between various quasi-subjects/quasi-objects, be they human or otherwise.

Such an approach may seem cold and dispassionate, but the aim of Bennett and Latour is the very opposite, wanting to centre affect and passion as fundamental forces operating *through* us and, indeed, sustaining our collective and individual assemblages. The power and agency of the student-chip-goblet-teacher assemblage is not ascribed to the whims of the student or the teacher and the passive obeisance of the goblet, or even to the goblet's resistance (because of its shape and colour) to the student's filling or the teacher's pointing.

Assemblages and processes of becoming

Body-assemblages are porous and partially closed, individuated only provisionally while their borders are sustained through various acts of embodiment. As we saw in the example in the previous section, mathematical concepts partake in this ongoing process. Among those acts of embodiment are the movements of the hand as it drops chips into a goblet, the placing and arranging of chips on the table, the use of the voice and gesture to bond with the other students, the teacher, etc. But we want to attend to how the 'space of joint activity' – where the students and teacher are physically arranged around the tables to collaborate – includes not just chips and containers but mathematical concepts, as well. Moreover, the 'psychical and social inscription' to which Grosz alludes is not simply imposed on the material conditions in the situation, but is in fact enlivened and articulated by the active force of these material conditions. In presenting this example, we aim to prompt discussions of how we determine boundaries around bodies and how these processes of determination inform our theories of learning.

As we saw in the earlier section on theories of embodiment, one major concern with such theories is that they tend to locate knowing in the individual body and do not adequately address the collective social body, which is a material network-body.[8] This is evident even in Radford and Roth's summoning of the collective space of joint activity. In other words, the body needs to be seen as part of a complex material network in which the social *and* the material are fused. Common sense tells us that the body is

[8] A network-body is a body connected and constituted through a rhizomatic lattice of material/social coupling. The rhizome metaphor has become an insightful way of conceptualizing complex interaction in the social sciences, including recent literature in education (de Freitas, 2012a; Gough, 2004; MacLure, 2010; Ringrose, 2010; Semetsky, 2006; Webb, 2008). A rhizome has no centre or primary root but grows and disperses itself through multiple entry and exit points.

an individual, discrete entity and that cognition occurs within its borders. Post-humanist theories of subjectivity, however, have shown how subjects are constituted as assemblages of dispersed social networks and have argued that the human body itself must be conceived in terms of malleable borders and distributed networks (Deleuze & Guattari, 1987; Bennett, 2010; Rotman, 2008; Latour, 2005). With such distributed networks, the concept of agency must also be rethought, because such an approach problematizes the premise that any one part of the assemblage is the source of action, intention or will. Such problematizing will mean revising notions such as student agency, as well as advocacy or interventions for improving or supporting student agency. We will need to reconceive agency as operating within the relations of an ever-changing assemblage, a force that flows across the encounters between goblet, chip, finger, voice and instruction, so that the very idea of the corporeal will become complicated:

> The emphasis on corporeality further dislocates agency as the property of a discrete, self-knowing subject inasmuch as the corpus is now recognized as exhibiting capacities that have significant effects on social and political situations. Thus bodies communicate with other bodies through their gestures and conduct to arouse visceral responses and prompt forms of judgement that do not necessarily pass through conscious awareness [...] Paying attention to corporeality as a practical and efficacious series of emergent capacities thus reveals both the materiality of agency and agentic properties inherent in nature itself. Both have important implications for the way we understand political processes. (Coole & Frost, 2010, p. 20)

The challenge is to begin to map the way these visceral responses build into blocks of sensation and to show how they are implicated in everyday classroom practices and thus have political impact. In Chapters 3 and 4, we study one facet of this complex intermingling of the micro and the macro, showing how theories of sense and perception were historically linked to an aestheticization of mathematics that influenced contemporary policy in mathematics curriculum and instruction. In Chapter 5, we unravel the materiality of language use in classrooms and show how movement and affect circulate alongside the spoken word. Various theoretical approaches to education try to attend more carefully to the collective emergent nature of learning, but they frequently fall back onto language that indicates their commitment to a theory of individuated being. For instance, even in Hutchins's (1995) ground-breaking study of distributed cognition, cognitive processes are shown to be distributed (as problems are solved on a ship), but they are seen as being distributed across members of the social group, who coordinate material and environmental structures. In this book, we argue that learners' and teachers' bodies are always in the process of becoming assemblages of diverse and dynamic materialities, including physical objects such as pencils, compasses and calculators, and also mathematical concepts. Perhaps the best way to think of an assemblage is to imagine a knot of many different threads, twisting and tangling, composed of loose run away strands and tight little balls of interwoven density. Knots have no inside or outside, no beginning or end – one is always in the middle of the knot, always moving along its various threads. The knotted assemblage is composed of diverse elements and vibrant materials of all kinds. In the case of the classroom, these knotted assemblages are composed of humans, writing implements, writing surfaces, texts, desks, doors, chips, as well as disciplinary forces and habits of control and capitulation. Assemblages are also saturated and sustained by the non-visible, be it sonic, haptic, olfactory or some other material percept. They are not simply cobbled together with discrete add-ons of parts, be they goblets, organs or concepts, but are always in motion, growing and contracting.

In other words, we study the body more in terms of its becoming than its being. Thus, rather than being self-contained, a body is a set of material relations that seems to structure the other material relations around it. These relations are themselves undergoing other changes in relation to other kinds of forces. Given that assemblages are primarily about relations, we take a body to be an 'extended and relatively closed system of differential relations' (Cutler & MacKenzie, 2011, p. 55). We use the term 'differential relation', which is *meant* to evoke the mathematical term used in calculus, to emphasize difference over identity in the quivering unstable assemblages that constitute bodies.[9] Focusing on difference always privileges change over stasis, becoming over being. How does the student-chip-goblet-teacher assemblage change as the chips are placed beside the goblets or as the space on the desk is exhausted? To centre the concept of difference, rather than identity, in a theory of embodiment is to point to the intrinsic indeterminacy and mobility at the heart of any process of becoming, which is crucial to a de-essentialising of the body. The term also points to *relations* as more ontologically primitive than the *relata* (or objects that are being

[9] Cutler and MacKenzie (2011) borrow this term from Deleuze, who nicely leverages this mathematical concept to talk about processes of determination in which relations are more ontologically primitive than relata.

related one to the other) that are determined through those relations, and it thereby fundamentally shifts our ways of thinking about matter. Thus, the child-cube-concept assemblage that we discussed in the previous section is no more than an open set of unstable relations, some of them more constraining or limiting than others. Assemblages are porous and partially closed systems that are never entirely cut off from that which surrounds them, but they are differentiated from their surroundings by degree, as though they were contractions of intensity or energy. The line that links the students to the chips is not fixed or rigid, and neither is the line that links the child to the cube. However, this line or relation with other materialities may temporarily fade. Once a particular practice is enacted, such as putting the chips beside the goblet, gesturing between them, describing the cube using words, or exploring the cube using hands, the bond begins to tighten, and the forces that modulate the bond are dampened, as language and limbs enter into the complex tangle. We offer the 1970s image of a lava lamp, in which there are temporary coagulations (assemblages) that morph over time, sometimes growing and sometimes falling away, but always moving under the influence of their surroundings.

In contrast to scholars such as Latour and Bennett, Ingold (2011) has argued that the notion of agency is unhelpful because it inevitably reinforces thinking in terms of identity and causality. Of course talk of agency arises naturally as we try to describe how humans can act, as well as how objects seem to 'act back', through resistance or accommodation. He proposes that we think of humans and objects as materials-in-motion onto which agency is ascribed. In other words, our language imputes agency when it is not actually there. He points out that the subject-verb structures of language (as in 'the wind blows') make it difficult to discuss air current without imputing agency. Similarly, it is hard to study the student, teacher or other 'individual', for example, in terms of a diffuse 'hive of activity' (p. 17). Thus, in shifting our attention to a distributed agency, we need to reckon with a dissolution of our language-use. In Deleuze's terminology, this is a turn towards the exteriority of thought, an attempt to map subjectivity as a process of becoming, as a mobility, rather than a fixed structure of thinking parts and feeling organs. Indeed, Rotman (2008) also draws on Deleuze in his overhauling of the concept of the body – and embodiment – in terms of distributed agency across a network of interactions, the properties of which are constantly changing. In other words, the body is no longer confined to the flesh borders of the individual person. Rotman's refrain of 'becoming beside ourselves' captures the new acentred sense of subjectivity that is emerging this century, in part because of new digital technologies that herald and hail a network 'I' which thinks of itself as permeated by other collectives and assemblages.

> Such an 'I' is immersive and gesturo-haptic, understanding itself as meaningful from without, an embodied agent increasingly defined by the networks threading through it, and experiencing itself (not withstanding the ubiquitous computer screen interface) as much through touch as vision, through tactile, gestural, and haptic means as it navigates itself through informational space, traversing a 'world of pervasive proximity' whose 'dominant sense is touch'. (de Kerckhove, 2006, p. 8, cited in Rotman, 2008, p. 8)

This 'I' is plural and distributed, 'spilling out of itself' while forming new assemblages and new folds within its tissue. Rotman also examines the practices of mathematicians, so that when he speaks of the network 'I', he is referring (also) to the calculations, symbols and theorems of mathematics, as well as to mathematicians themselves. He sees in digital technologies openings to the rather tight networks that have dominated the discipline, arguing that the temporal and spatial expressions of mathematics will profoundly affect what mathematicians do and what it means to live a mathematical life (Hersh & John-Steiner, 2011). We will pursue this theme in some detail in later chapters, where we examine more closely the role of sense and sensation, as well as technology-enhanced instruction, in mathematics classrooms.

The theory of assemblage we are proposing in this book allows us to reimagine agency as distributed across the surface of these heterogeneous alliances. This theory suggests that the power of an alliance is mostly enhanced by the heterogeneity of that alliance:

> Bodies enhance their power in or as a heterogeneous assemblage. What this suggests for the concept of agency is that the efficacy or effectivity to which that term has traditionally referred becomes distributed across an ontologically heterogeneous field, rather than being a capacity localized in a human body or in a collective produced (only) by human efforts. (Bennett, 2010, p. 23)

Power is not distributed evenly across the surface of an assemblage, because there are joints or nodes where there is more traffic and affect than at others. Assemblages have 'uneven topographies' and possess emergent properties (Bennett, 2010, p. 24). Certain acts of individuation, often involving speaking or becoming visible, leverage this power differential. The study of assemblages demands careful attention to the way that affect or feelings emerge and are mobilized or blocked. It is important to conceive of the agency of the assemblage in terms that do not simply reduce it to a static structure imposing fixity on the active agents within it. The power of the assemblage is not merely negative as a constraint or passive as an enabler.

Although this chapter began with the human body, we have now moved to a more general focus on materiality and becoming. This move to materiality is crucial, we believe, if theories of embodiment are to adequately address the nature of mathematics. In rethinking the contours of the human body and shifting our attention to processes of becoming, we now have to grapple with the way that mathematical concepts partake of those processes. In the next chapter, we begin to articulate a new kind of materialism that aims to do so. Like work in complexity theory (Davis & Simmt, 2003), our approach rests on a radical rethinking of matter itself – something that will allow us to redefine mathematical agency as being fundamentally relational. We explain how our new kind of materialism – which we name *inclusive materialism* – offers a productive framework for studying mathematical activity and mathematics education. We draw on recent work in the philosophy of science, focusing in particular on work which studies the implications of the intrinsic indeterminism of matter (Barad, 2007, 2010, 2011; Coole & Frost, 2010).

2 The 'ontological turn' of inclusive materialism

In Chapter 1, we discussed a new way of thinking about bodies as assemblages. In this chapter, we extend our proposal to thinking about matter more generally and argue that this new approach lends itself to the study of mathematical activity and mathematics education. Questions about the nature of matter are usually deemed ontological, but such questions also inform our beliefs about life more generally and are thus implicated in our understanding of identity, culture and education. We introduce a philosophical framework – *inclusive materialism* – and show how this framework allows us to theorize mathematical concepts in new ways. We begin this chapter with a brief description of how inclusive materialism relates to other kinds of materialism in the history of philosophy. We outline Karen Barad's 'agential realism', which challenges classical ontologies of mind and matter and is based primarily on her interpretations of quantum physics. Her ontological shift has major consequences for the way we think about the human body, but also for the way we think about theoretical concepts. We describe in some depth the physics experiments that have contributed so profoundly to Barad's perspective and begin to explore the ways in which her interpretations of these experiments might bear on mathematics education. We see our work as complementing that of the scholars who emphasize mathematics as a historically and culturally rooted domain of embodied activity.

New materialisms

The term *materialism* has a long history in philosophy, as well as in political theory. In philosophy, materialism has been used to describe a kind of ontological monism that declares all that exists to be matter. Materialism has thus been aligned with various kinds of empiricist epistemologies that centre the senses and sensation in our coming to know. The question as to what constitutes matter, however, has been answered very differently by different materialists. Baruch Spinoza invested matter with immanent vitalism, whereas David Hume imagined matter as mechanistic. In the nineteenth century, the term 'historical materialism' was used by Marxist theorists to describe the socio-economic force of everyday material conditions in shaping power structures. It is this complex history of the term – its political, ontological and epistemological heritage – which makes it a powerful tool for examining experiences of mathematics education.

If the sociocultural or linguistic turns in the study of learning saw the demise of previous materialisms, a new materialism must incorporate the important insights from these discursive approaches and be wary of returning to a naïve or crude empiricism that denies these insights. Indeed, such a history must be reckoned with, especially now, if we are to think our way through the new neurobiological accounts of learning to a space where there is ample room for complex understandings of agency.

Many of our received ideas about materiality can be traced to Descartes' definition of matter as essentially 'length, breadth and thickness; as extended, uniform, and inert' (Coole & Frost, 2010, p. 7). Such an image of matter lent itself to quantification, calculation and measurement. According to this model, the actions of bodies are entirely mechanistic and determined by physical laws. Of course, Descartes and other Cartesians saw human bodies as exceptional in their freedom, will and mindfulness. The human body is occupied by a human mind (and soul) that, according to the Cartesian reading, animates and mobilizes it, while other bodies lack any such mind. New materialisms critique this image of modernity, arguing that freedom and agency must be seen as dispersed across human and non-human agents, thereby troubling any ontologies that deny mobility and action to seemingly inert matter or immaterial concepts (Latour, 2005). New materialisms argue that theories of subjectivity and the social will be augmented by a new theory of matter that resists the binary divide between human agency and inert, passive matter (Barad, 2003, 2007; Latour, 2005; Malafouris, 2008). While avoiding the dangers of a naïve vitalism, new materialisms break with binaries that set organic against inorganic – and animate against inanimate – so that matter might be reanimated more generally and seen in terms of potentiality and emergent generative power. New materialists often 'eschew the distinction between organic, inorganic, or animate and inanimate' and discern 'emergent, generative powers (or agentic capacities)' within matter more generally (Coole & Frost, 2010, p. 9). For those working from a feminist perspective, new materialisms seek to study the bond between discursive and non-discursive forces in new ways, while sustaining the insights gained from previous work in this area. The body is no longer simply passive matter upon which culture is inscribed, nor is it treated as a mute, precognitive container for consciousness; rather, it is centrally implicated in the constitution of itself and society. New materialisms rescue the body, so to speak, from a theory of discourse that denies its materiality in order to grant the body some measure of agency and power in the making of subjectivity.

These new ways of thinking about matter are relevant to our experiences of subjectivity and culture. At the same time, these new materialisms render previous distinctions untenable, forcing us to reconsider the ways we demote the non-sentient in relation to the human and troubling the normative boundaries that ground much of our thinking about politics and ethics. Any materialist analysis of relationships of power and subjection within education has to attend not only to the enduring political and economic structures that operate through bodies, but also grapple with the diverse scales and temporalities at play, and also identify the vulnerabilities and contingencies of the visceral, and even molecular, entanglements that sustain and alter these structures. Work in cultural-historical activity theory moves in this direction as well, at least with regard to the first point. For instance, Radford (2008) suggests that 'subjectivity and cultural objectivity mutually overlap and where the mind extends itself beyond the skin' (p. 219), and more recently in 2012 he argued for 'a monistic position according to which mind is a property of matter. More specifically, mind is conceptualized as a feature of living material bodies characterized by a capacity for responsive sensation' (p. 2). What we do not see in the current literature on the material aspects of mathematics, however, and what we aim to do here, is to show how mathematical concepts partake of the material in operative, agential ways. In much of the literature on embodiment (discussed in Chapter 1), we find that learning is assumed to have a teleological trajectory towards fixed and immovable mathematical concepts. Concepts are said to emerge through activity, but there is no troubling of the specificity of the concepts – in other words, the mathematical concepts (multiplication, cube, zero) are taken for granted, while students collaboratively move towards them. In this we see a continued, if tacit, belief in what Wittgenstein (1978) called 'this glitter of the concepts' (p. 274). In this chapter, we work towards a new ontology of mathematics that can help us study the materiality of mathematical concepts (to move away from the glitter and obscurity, as Wittgenstein hoped to do), and we argue that doing so allows us to rethink the nature of learning, teaching and doing mathematics.

We propose an *inclusive materialism* that emphasizes the biopolitics of all phenomenological studies, with the aim of mapping the links between the micro-visceral activity of the body and the political forces that flow across and through these bodies. In Chapters 6 and 7, for instance, we map the intricate interlocking (imbrication) of everyday corporeal practices in classrooms with regional and global markets of mathematics curriculum policy. Our hope is that inclusive materialism moves away from a theory of power as a totalizing, external force and follows power as it flows through sensation and affect, across the surfaces of bodies as they emerge in relation to these flows. We consider the work of Barad and other related scholars to be post-humanist in its turning away from humanist conceptions of agency and rationality, while remaining committed to mapping the political configurations of life. This renewed attention to a critical agenda builds on the work of Latour and Malafouris and their commitment to attend to various kinds of agencies that have been denied value or status in the past. Our aim is to combine methods from phenomenology and perception studies with new feminist philosophies of science that frame such work in socio-political terms. As Deleuze and Guattari (1987) suggest, 'everything is political, but every politics is simultaneously a *macropolitics* and a *micropolitics*' (p. 213, *emphasis in original*).

For many educational researchers working within a socio-political paradigm, however, references to nature, matter and the behaviour of non-human animals has been seen as the mark of a positivist model of research. The social sciences emerged from the physical sciences in the nineteenth century, but reliance on principles and methods from the physical sciences has been problematic when pursued under a positivist paradigm, as in the infamous cases of sociobiology and eugenics. On the other hand, such reliance is considered entirely par for the course by many researchers who apply models and methods from laboratory science to case studies of human interaction. What we find inspiring about Barad's work on quantum physics, which we discuss in the next section, and other recent work on bioethics and biopolitics (Bennett, 2010; Kirby, 2011), is the way this work references contemporary scientific developments that go against the grain of positivism. Not only is the science different, operating with different ontological and epistemological commitments, but the applications of this science to the social are also entirely different, serving a political agenda that aims to address the unexamined beliefs that are at work in the structuring of injustice and inequity. As Coole and Frost (2010) suggest, 'The great materialist philosophies of the nineteenth century, notably those of Marx, Nietzsche, and Freud, were themselves hugely influenced by developments in the natural sciences, yet the new physics and biology make it impossible to understand matter any longer in ways that were inspired by classical science' (p. 5). The twentieth-century sciences of chaos, complexity and string theory, among other developments in biology and chemistry, have revolutionized our ideas of causality, matter and life. As Gleick (1987) remarks, 'To some physicists chaos is a science of process rather than state; of becoming rather than being' (p. 5). For contemporary scientists, the physical world is conceptualized as a dynamic and nonlinear system 'exhibiting immanently self-organizing properties' (Coole & Frost, 2010, p. 13). The focus of this new emergent paradigm in science seems to be on complex processes of transformation and reconfiguration, mapping intensity and energy rather than seeking states of equilibrium.

We want to emphasize how inclusive materialism is different from another kind of contemporary materialism sometimes labelled 'neo-Darwinian materialism', which the philosopher Thomas Nagel critiqued in his book *Mind and cosmos* (2012). Indeed, we suspect that the debate sparked by the publication of that book points to the need to rethink materialism in new terms. It is worth outlining four important aspects of inclusive materialism that distinguish it from other related approaches.[1] First, we do not want to *reduce* the universe to one kind of substance. While we are working with what Delanda and others call a flat ontology, insofar as we deny the duality of mind and matter, that does not mean we pursue *reductive* explanations. Rather than showing how all difference is reducible to some underlying sameness and how all materiality is determined through mechanistic causes, we pursue descriptions of phenomena that privilege difference and multiplicity, underscoring the virtual dimension of matter. Second, rather than remaining silent on political facets of experience, we study the way that material practices shape and are shaped by socio-political concerns. In fact, our aim is to show how the political and the material are inextricably entangled and to do so with the explicit concern of addressing issues of inequity in education. Our focus on bodily processes and corporeal capacities – even as we reconceive the body – locates our work within these larger concerns. Third, rather than explain away emotion and affect as secondary effects of more fundamental material entities and actions – or perhaps as irrational – we centre affect and aesthetics as engines of mathematical activity. Our discussion of affect in chapters 5, 6 and 7 would no

[1] To be fair, this popular conception is debatable. But the important point here is that we Feser (2010) has critiqued neo-Darwinian materialism. However, many would contest the accuracy of his description, especially those who are on the other side of the debate, such as philosopher Daniel Dennett and biologist Richard Dawkins (Ferguson, 2013).

doubt frustrate certain kinds of materialists because we refuse to privilege rationality: We explore the force of non-sense and paradox not as delusions, but as pivotal facets of mathematical activity. Moreover, we are keen to unpack constructs like mathematical ability and mathematical intuition, so that we might begin to study the way that mathematics functions in the making of 'common sense' in schools.

Lastly, we are also keen to recognize the astonishment, surprise and creativity of mathematical activity not merely as psychological dimensions of such activity, but as affects that are linked to the virtuality and vitality of matter. This kind of revitalising of matter would seem mystifying to more positivist brands of materialism.[2] In fact, our attempt to decentre human agency and study indeterminate processes of becoming would no doubt be considered a sort of dangerous posthumanism by many scholars who would otherwise align with our focus on materiality. For obvious reasons, there is something terribly uncomfortable about decentring human agency. But when we claim that inclusive materialism is post-humanist, it is important to keep in mind that we use the term to define a theoretical position that contrasts with humanist assumptions about the self, identity and humanity. We are not anti-human, as is evident in the socio-political commitments of our inclusive materialism, but we do want to unsettle the category of human (and life) in the same way that feminist, queer and critical race philosophers did in their attempts to expand inclusive practices. During the last few years, various kinds of posthumanism have proliferated, asking readers to rethink the way we sort life experiences into the categories of human and non-human.[3] We see our work as springing forth from these new branches of theory.

Towards a relational ontology

Barad (2003, 2007, 2010, 2011) has spearheaded much of the work in this new direction, using the work of Niels Bohr and quantum theory to argue for a dynamism or indeterminism that is inherent to matter. In order to better grasp the spread of agency across human and non-human agents, thereby granting matter and materiality the power to exceed discursive framing, Barad proposes a 'relational ontology' and 'agential realism' (p. 829). These terms are rooted in Bohr's work in quantum physics, which

[2] We discuss virtuality in Chapter 8.

[3] See for instance, the *Posthumanities Series* of books published by the University of Minnesota Press (which included twenty books as of 2012).

shows that atoms possess an inherent ontological indeterminism or lack of identity; that is, they possess an intrinsic duality in being both particle and wave. In other words, their identity as either particle or wave is something that is performed in concert with other entities (such as the measuring devices that are used to determine their status as particle or wave), rather than given *a priori*. Barad uses this work as a philosophical basis for talking about matter more generally, not just about the atoms, photons and electrons with which Bohr was concerned, but also about lightning, animals, plants and humans (see Barad, 2011). It is worth briefly describing Bohr's work, as it serves as an important 'generic' example for thinking about how Barad wants to rethink matter, agency and ontology.

Along with many other scientists, Bohr was involved in trying to determine the nature of subatomic entities. The double-slit experiment figured prominently in this venture. The experiment entailed shooting electrons at a screen with two parallel slits and studying the pattern that resulted on a second screen behind the first. If the electrons behaved like particles, the resultant pattern should be a scatter pattern, with an intense band of impact directly behind one or both slits. Instead the pattern observed was a series of parallel bands just like a wave diffraction pattern. This result seemed to suggest that the electrons were behaving like waves, entering the two slits and then interfering with each other on the other side of the first screen, either cancelling each other out or amplifying each other, depending on the kind of wave interference, before hitting the second screen. In an attempt to avoid the interference, physicists shot one electron at a time, and yet the same shocking pattern occurred. How could only one electron produce a wave diffraction pattern? The only conclusion was that each electron was both particle and wave, leaving as a particle but becoming a wave of potential that enters both slits and then *interferes with itself* on the other side! In fact, all the possibilities – one slit, two slits or none – must be considered in superposition. Einstein, who refused to accept Bohr's commitment to ontological indeterminacy, proposed creating a 'which-slit' detector that could reveal how the electron behaves when it gets to the first screen. Eventually, this experiment revealed a fascinating outcome. When an electron passes through the two-slit device with the which-slit detector in place, the pattern produced is that of a particle, and yet without the which-slit detector, the pattern is that of a wave. Bohr predicted this phenomenon and explained it in terms of '*the quantum entanglements* of the measurement apparatus and the object of measurement', emphasizing that 'it is not so much the case that things behave differently when measured differently; rather, the point is that there is only the phenomena – the intra-action of "apparatus" and "object" in their inseparability' (Barad, 2011, p. 143, *emphasis in original*).[4]

The philosophical consequences of Bohr's work pertain not simply to the nature of matter, but also to the nature of our measuring devices and discourse more generally. If we follow these consequences to their fullest, we will find that the ontological relationship between discourse and matter is fundamentally altered. It is not simply the case that our knowledge of the electron is uncertain, dependent as it is on human decisions. That kind of emphasis on epistemic uncertainty was Heisenberg's take on the contradictions of the two-slit experiment. His 'uncertainty principle' states that knowers interfere with that which they aim to understand. Quantum theory, however, is not simply about the relativism implied by Heisenberg. Rather, it proposes there are inherent indeterminacies in matter. In other words, Bohr's quantum theory makes an ontological claim, not (only) an epistemological one. He explains that indeterminacy is intrinsic to matter and not simply a reflection of our limited skills at observation and measurement. He recognizes the way that meaning and matter are mutually entailed: Things do not have determinate boundaries, and properties and words do not have inherently determinate meanings:

> The objective referent for concepts like 'wave' and 'particle' is not a determinately bounded object with inherent characteristics [...] but rather what is called a *phenomenon* – the entanglement/inseparability of 'objects' and 'apparatus' (which do not preexist the experiment but rather emerge from it). (Barad, 2011, p. 142, *emphasis in original*)

Barad (through Bohr) adopts a quantum ontology in which meaning and matter are bound and woven together in a kind of 'onto-epistemology', thus overcoming the classical approach in which ontology precedes epistemology. In rejecting the separability of knower and known, Bohr denies that concepts are 'ideational' or immaterial and claims instead that

[4] Not only does this support Heisenberg's uncertainty principle, but it also demands that we think about observation in radical new terms. What exactly constitutes an observation? Is observation peculiar to humans or human-made devices? Is it just an act of recording information? Would a watching dog or other animal interfere, as well? How might this line of questioning perturb mathematical Platonists when they make claims, such as the one Hardy (1929) made: 'I have myself always thought of a mathematician as an observer, a man who gazes at a distant range of mountains and notes down his observations' (p. 18).

concepts are specific physical arrangements (Barad, 2003, p. 814). He would thus also reject the notion of 'mediation', a term that often gets used in describing the relationship between mathematical concepts and devices (apparatuses) as well as between mathematical concepts and the human body. Mediation implies pre-given substances and definitive positions that pre-exist the relations between them. Processes of mediation tend to assume that devices and bodies work in the service of acquiring a concept and, as a result, such devices and bodies are ultimately seen as optional or extraneous to that concept. Think, for example, of the perceived relationship between a circle and a compass – is the compass optional to the meaning of 'circle' or at least discardable after a suitable amount of time? The idea that the concept of circle is mediated by the compass fails to grasp fully their mutual entailment, nor does it appreciate the process through which 'circle' and 'compass' are provisionally assembled together in a phenomenon. As we will see, the same can be said for the suggestion that the circle is mediated by language or mediated by the thumb-to-index-finger rounded gesture. In Chapter 5, we use the term 'modulation', instead of mediation, to get at the complex ontological entanglement of language and body.

Barad introduces new terminology that might help us rethink the ontology of mathematics; she uses the term *intra-action* to speak of the relationships that bring about the distinctions between apparati and objects and, more generally, between mattering and meaning. This term signals her move away from a concern for the way two distinct realms (words and things) *interact* and towards an understanding of what is happening between two mutually entailed folds of the same realm. The atom and the two-slit device intra-act, as do the circle and the compass. But perhaps more significantly, these two entities are always in a process of becoming, a process that entails diverse agencies. Our approach is similar to that of Barad in that we decentre the human subject as the wilful agent, and we animate the mathematical concept, so that 'interaction' between the two becomes intra-action. We find this approach particularly relevant to discussions of mathematics, given that mathematical concepts are all too often seen as inert, inanimate and immaterial. Barad's new materialism assumes that theoretical concepts are like everything else, insofar as they exist without determinate boundaries. One consequence of intra-action is the inseparability of concept and matter. Another key consequence, which we noted in passing earlier in this section, is that of performativity. It is central to Barad because it moves away from an essentialised notion of identity and towards a more temporal and contingent sense of becoming. Barad expands on previous theories of performativity that emphasized discursive power, so that the body and the physical world are seen to partake in this intra-active performance. Indeed, Barad has described her overall project as a 'diffractive' reading of the insights from both physics and feminist philosophy, which enables her to recast performativity as a facet of new materialism.

Such an approach challenges the classical ontological position in which the 'world is populated by independently existing things with determinate boundaries and properties that move around in a container called "space" in step with a linear sequence of moments called "time"' (Barad, 2011, p. 144). Indeed, Barad argues that rather than existing as determinate givens, both space and time emerge intra-actively in the making of phenomena. Since Einstein, we have become accustomed to conceive of space and time as being dependent on each other – that is, as space and time interacting – but Barad's intra-action dissolves any pre-given boundaries between the two, so that *spacetimemattering* is a phenomenon. This phenomenon can be thought of as the tangled mesh of spacetimemattering in which the past is not some known fixed location given by a space-time coordinate and the future is not simply a point that marks the endpoint of a trajectory.

Barad's assertions may seem overly dependent on the astonishing behaviour of subatomic matter to warrant the interest of mathematics educators. Barad (2011) has also explored situations other than the subatomic in which these assumptions seem to be called into question, where classical conceptions of causality and identity fail to explain particular complex performances by, for example, lightning bolts, stingrays and humans,[5] but her focus on quantum physics highlights space-time oddities in which, for example, 'communication' occurs with neither a sender nor a recipient until the message has already been 'transmitted', while 'movement' occurs in quantum leaps so that an entity can go from one energy level to another without having been anywhere in between. Classical ontology fails to account for these situations, whereas Barad's relational ontology succeeds. At the same time, classical mechanics manages to explain very well certain regularities in the behaviour of atoms, as well as lightning, and it clearly resonates quite well with how we go about our everyday activities.

[5] Barad (2011) describes these oddities as 'queer critters', evoking not only the uncanny nature of the examples she describes, but also acknowledging her debt to the performativity work of queer theory, as well as her political agenda of being attentive to the activities of animate and inanimate beings.

And yet, as Coole and Frost (2010) suggest, Barad helps us rethink the 'real' on larger scales:

> The atom is a smeared field of distributed charge whose subatomic particles are less like planets in solar orbit than they are like flashes of charge that emerge from and dissipate in the empty space from which they are composed. Even when vast numbers of atoms are assembled in the kind of macrostructures we experience in the 'condensed matter' of the perceptible world, their subatomic behavior consists in the constant emergence, attraction, repulsion, fluctuation, and shifting of nodes of charge: which is to say that they demonstrate none of the comforting stability or solidity we take for granted. While this does not of course mean that the objective world we inhabit is mere illusion, it does suggest that even – or especially – the most ardent realist must concede that the empirical realm we stumble around in does not capture the truth or essence of matter in any ultimate sense and that matter is thus amenable to some new conceptions that differ from those upon which we habitually rely. (Coole & Frost, 2010, p. 11).

Barad's (2008) quantum-inflected new materialism claims that discourse and matter are literally entangled and mutually entailed. She makes this argument with reference to the famous thought experiment of Schrödinger's cat (in which the cat is penned up in a steel chamber along with a tiny bit of radioactive substance that may or may not decay with equal probability within the hour). If one atom of the substance decays, the Geiger counter then discharges and releases a hammer that shatters a small flask of hydrocyanic acid, killing the cat. The fates of the atom and the cat are entangled. We cannot predict with certainty whether the atom will or will not decay, but we can calculate the probability that an atom will decay in a given time period. Probability plays a central role in producing a system with a dead-*and*-alive cat 'smeared out in equal parts' (Schrödinger, 1935/1983). Rather than simply inferring from this paradox that we are epistemically limited in comprehending the fate of the cat (as Heisenberg had when he interpreted the two-slit experiment as an epistemological human weakness), Barad argues that the cat and the atom do not possess 'determinate states of existence' – there is no fact of the matter regarding the cat being dead or alive without our measuring it. The actual material existence of the cat and our observing the cat are ontologically entangled events. Moreover, the very concept of 'life state' is no longer well-defined because it is without determinate meaning.

The thought experiment, which is itself an assemblage of material mechanistic devices and instruments that measure or code the phenomenon under study, forces us to reckon with the ontological implications of our *knowing* something. The paradox perturbs our understanding of how to assign the words 'alive' or 'dead' to the cat. The power of the thought experiment rests in part on its troubling of our ability to determine whether the animal is alive or dead – which by all counts is considered a rather important ability – and thus it underscores our need for a more robust epistem-ontology. Barad's (2008) move from *interaction* between two distinct realms (words and things) to their *intra-action* stems from the fact that the discursive boundary determining 'alive' and 'dead' is tangled up in the materialities of the cat, the atoms and the Geiger counter. The point is not that words function within language games and that "alive" and "dead" are provisionally defined through sociocultural forces. This would miss the materialist and ontological argument altogether. Rather, the point is that there is some *fact to the matter* of this strange cat-atom assemblage and its indeterminate existence.

This is an attempt to grapple with the consequences of a non-representational theory of language, to grasp the gritty materiality of discourse in practice. The entanglement of the discursive and the material implies continuously mutating and enacting boundaries so that words and things become *provisionally* construed as mutually distinct. Individuations occur through boundary-making practices whereby the 'superposition' of contraries, or mutually distinct entities, such as dead or alive, are seemingly cut out of the fold and separated (p. 173). Language performs through a material reconfiguring of the world, rather than through replicating, modelling or coding a material entity outside of it.

> Materiality is a discursive performance of the world, but discursive practices are not reducible to human-based actions. Matter does not serve as a mere support for discourse, nor is it merely the end product of human-based citational practices. Rather, discursive practices are specific material configurations/(re)configurings of the world through which local determinations of boundaries, properties, and meanings are differentially enacted. That is, discursive practices are ongoing agential intra-actions of the world. (p. 173)

Although Barad speaks of materiality as a discursive performance, which might imply a reducing of materiality to discourse, she is adamant about the materiality of discourse itself, hence the word/thing intra-action. Phenomena come to matter through discursive practices, but discursive practices are always already material. Matter and meaning are inseparable and do not stand in a relation of exteriority to each other. Such a claim may initially seem non-operational, insofar as it denies any theory of reference between the two domains, but one can begin to pursue this vision through experiments in analysis, describing and studying activity for evidence of this entanglement. We perform such experiments throughout this book, demonstrating how one can study various facets of mathematics education according to this point of view. This is akin to the kind of thought experiment that Bohr pursues by helping readers begin to imagine the consequences of his theoretical position. In the next section, we begin to pursue the implications of inclusive materialism in relation to teaching and learning mathematics.

The pedagogy of the concept

We turned to Barad's relational, or quantum, ontology because we were seeking ways of explaining certain complex questions about the way the socioculturally immersed body is implicated in the knowing and doing of mathematics. These questions warranted for us a rethinking of the very notion of a discrete, independent body that simply interacts with mathematical concepts. Barad's quantum ontology supports our interest in assemblages, allowing us to move beyond a theory of mediation that treats, for instance (with reference to the example in Chapter 1), the child and the cube as independent, pre-known entities. Barad's revisions of space and time are subtler, but they enable us to approach notions of agency and causality differently, which impacts the way we think about learning. Rather than study learning in terms of a classical ontology, in which agency is reduced to mechanistic models of cause and effect, Barad's relational ontology allows us to map the complex entanglement of learning and becoming. What might be the boundary-making practices that are involved in the emergence of mathematical concepts? What kinds of cuts or articulations *perform* abstract, immaterial entities?

Consider, for instance, a diagram generated by a grade seven student when asked to solve the following problem: 'At every corner of a square fishpond there is a tree. Make the fishpond twice as big so that it remains a square but the trees remain where they are' (see Figure 2.1).[6]

The diagram fuses the symbolic square with the iconic images of the trees, flattening them onto the page in such a way that no single point of view could be said to be conveyed. This conjunction of different registers – in this case, the symbolic and the iconic – is a powerful instrument in

[6] See de Freitas and Zolkower (2011) for a detailed analysis of student and teacher engagement with this problem.

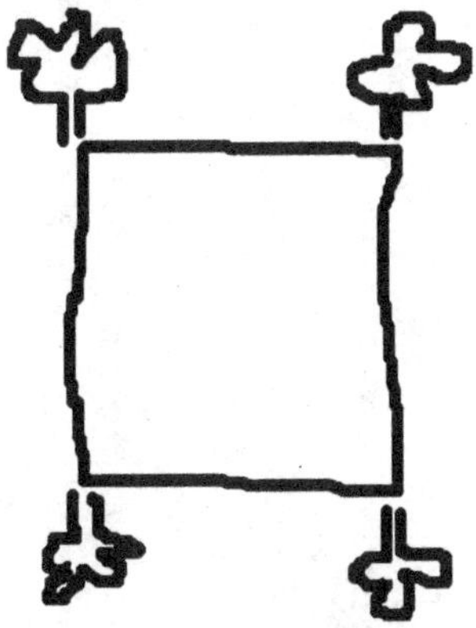

Figure 2.1. Four trees around the fishpond.

problem solving. In bringing together the trees and the square and the disembodied perspective, this simple act of inscription is a boundary-making practice, a device that reconfigures the world rather than simply representing it or coding it. The drawn lines both conjoin and separate the 'real' from the mathematical, the matter from the meaning. In Barad's terms, this is a discursive practice that actively engages with matter-meaning while enacting a cut that also inevitably separates the square-tree assemblage from, say, the fishpond as it might be imagined or remembered by the students. The trees are not *only* iconic representations of 'real' trees; they are also part of a diagrammatic device that effects the individuation of 'real' trees, recombining the material world with the new material square and simultaneously engendering the meaning of square. This device begins to perform the *meaning* of square and, as the children engage with it, the apparatus enacts this splitting. Thus, the diagramming is not merely an act of meaning-making, but rather it engages with and reconfigures the imbrication of meaning and matter. In other words, meaning and matter are intricately recombined through this simple sketch.

In Figure 2.2, generated collaboratively by the students and teacher, the diagram grows into a new assemblage incorporating new objects, such as dotted lines, diagonals and ninety-degree angles – a new configuring of the meaning of square and the mattering of trees. But the funny little trees persist, presented from a flattened, disembodied point of view, and this particular part of the discursive practice is crucial in *performing* the boundary between matter (in this case, trees and pond) and meaning (in this case, the square). The way these semi-iconic trees break with perception and remain coupled to the diagram is crucial in performing the separateness

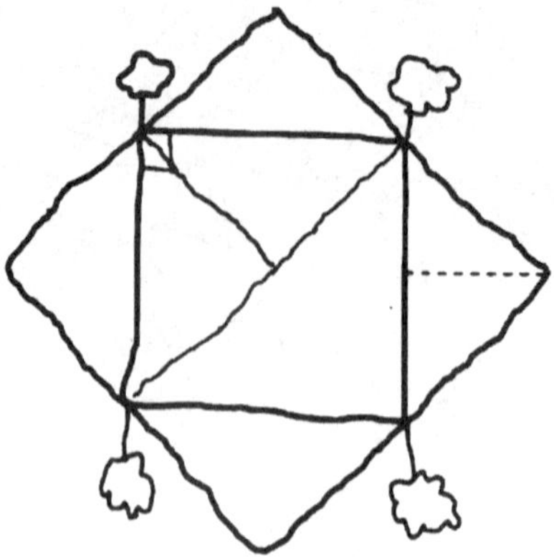

Figure 2.2. A reconfigured diagram with added elements.

of mathematical meaning. It is not the case, however, that the meaning *is* actually separate from the material world, but rather that the act of diagramming (like other boundary-making acts) creates a new fold or crease in the fabric of matter-meaning, so that we begin to operate as though there were two ontologically separate worlds.

Within the context of a mathematics problem, this example has highlighted the way in which we can challenge the reading of diagrams as representations of 'real world' or 'mathematical' things that remain separate entities. We begin to grasp how discursive practices are not ontologically prior to matter, because 'the relationship between the material and the discursive is one of mutual entailment' (Barad, 2007, p. 822). Any activity (and there is always activity) makes a 'cut' (making of boundaries) that configures one thing as distinct from another. As we discussed in Chapter 1, this directs our attention to processes of individuation, so that we might better recognize how the individual is not the starting point, but is provisionally emergent in relation to boundary-making practices that, for example, distinguish the animate child from the inanimate square or the child's cultural gender from his or her biological sex. Different 'cuts' might distinguish different assemblages, each with differing degrees of agency.

How might an inclusive materialism allow us to rethink learning as an indeterminate act of assembling various kinds of agencies rather than a trajectory that ends in the acquiring of fixed objects of knowledge? Can we make room within the epistemological project of mathematics education for the ontological engine of learning, namely the body in/of mathematics? How might the mathematics curriculum become more broadly concerned with the systematic seeking out and cultivating of vitality? How do we recast the curriculum to be vibrant and animate? We want to address these questions by focusing on what is typically considered the central goal of mathematics education – the development of conceptual knowledge – and its underlying vehicle: the mathematical concept. In the last section, we argued that the mathematical concept should not be considered an abstract universal detached from the particularities of its concrete instances. Rather than reducing concepts to immaterial, logical constraints, we follow Barad in arguing that concepts and matter are imbricated or interlocked. Barad claims that concepts are actually physical arrangements – where the 'physical' is a vibrant matter rather than an inert object. We propose a *pedagogy of the concept* that shifts the emphasis of theories of learning from the epistemological to the ontological, a shift that directs our attention to the materiality of mathematical knowledge:

> Perhaps the real challenge to Cartesianism lies not in reconfiguring how we know what we know but in dethroning the epistemological project itself by considering the body of knowledge on the same material plane of existence as the lived and the physical bodies? Perhaps the challenge is to treat learning as an ontological rather than an epistemological problem? (Cutler & MacKenzie, 2011, p. 63)

A pedagogy of the concept aims to revise concepts as material and ontogenetic devices. A concept, according to Cutler and Mackenzie (2011), brings with it an entire 'plane of immanence' on which we – as catalysers of concepts, which in turn catalyse us – operate, explore and create. Concepts are mobilized on this plane of immanence, which is often the material surface of encounter where the virtual is actualized. Indeed, the mobility of concepts grants them their creative power. This mobility is precisely what Châtelet will reveal to us in the next two chapters, as we consider his analysis of the history of mathematics. In a compelling inversion of the usual aim for learning, Cutler and Mackenzie suggest that learning is that which sustains the mobility of concepts, in other words 'that which resists determination as knowledge' (2011, p. 68). The pedagogy of the concept echoes Barad's commitment to the mutual entailment of matter and meaning and is meant to sustain student awareness of the vitality and indeterminism that lies at the source of curricular concepts.

Consider, as an example, the concept of the number line, which exemplifies the very idea of extension. As discussed earlier, Descartes considered extension the one defining quality of matter to which all other qualities, including motion, are derivative. Disrupting this Cartesian image of matter involves rethinking extension as somehow made muscular through the mobility of matter. Châtelet contrasts the lateral stretch of extension with the cutting and folding of new virtual dimensions through 'intension'. In this, he follows the scholastic tradition of contrasting extension – the interval actually travelled and its duration in time – with intension – its quickness, slowness or 'lateness' (Châtelet, 1993/2000, p. 38). As odd as this distinction might seem to modern readers, it is used by Châtelet to disrupt the privileging of position over motion and to try to imagine motion as the ontogenetic force by which position (or extension) comes into being. These new virtual dimensions are a kind of intensity that deforms the linearity of extension. The intensity of the new vertical dimension is what makes extension plastic and elastic. As we discuss more carefully in the next two chapters and also in Chapter 8, Châtelet uses the 'virtual' to describe the 'indeterminate dimension' in matter that literally destabilizes the rigidity of extension (p. 20). He describes these elastic folds in terms of transverse and vertical impulses that push through the apparent rigidity of extension.

In rethinking the number line, we might want to recast it as elastic, so that students can carve out the virtual real numbers embedded between whole numbers by grabbing and stretching the number line in order to bring forth an infinitude of numbers that were imperceptible a moment earlier. Students can then explore the density of the number line through this engagement – an engagement made all the more embodied when we see how an elastic line, more so than a rigid one, seems to sustain the potential of our flowing gestures. The number line is the quintessential form of extension, and thus it is crucial to see how this example helps us rethink extension (and the related concept of dimension) as a quivering space of potential rather than an interval contained between two fixed points. Through the eruptions of intensity, the concept of extension is 'charged with tensions and reveals itself as the positive condition for the birth of structure' (p. 105). The number line becomes a highly animate concept made vibrant and creative through the indeterminacy buried in it. The body of the number line engages with the body of the student, and a new kind of body-assemblage comes into being. We can see in this example the power of digital dynamic diagrams to afford the students this kind of opportunity.

Consider another concept from the elementary curriculum: the concept of zero. We can recast zero as the bringing into existence of nothing; the flight into integers; the excavation into rational numbers. A curriculum that is less concerned with procedures and more concerned with creativity might delay the upward extension of the counting numbers and the early practice of addition-as-grouping, and first make time for children to engage with the new virtual spaces created by the concept of zero. One can ask: What's the mobilizing virtuality of the concept of zero? How does it get massacred in the curriculum? In some textbooks, zero is introduced by the recursive subtracting of one or by counting down. So students are invited to imagine five objects, then four, then three, then two, then one, then zero. In this stepwise descent, zero is squarely placed in the real terrain of taking away the last remaining object, and we can hear the teacher saying, 'It's like eating the last cookie or using the last tissue.' But this kind of activity seems to designate zero as a lack, or absence, rather than as a generative concept.

Instead of being the nothing that is arrived at after subtraction, zero can be the motor by which motion arises. As a motor or a source of potentiality, zero is cast in a very different light, one that invites the placing of the index finger or the pencil tip on the paper: 'Here begins zero.' The point does not merely represent zero; it is the starting point out of which motion will produce magnitude and eventually roll over the counting numbers. Furthermore, as Châtelet argues, zero thus conceptualized marks a symmetry of choices towards the positive or negative magnitudes. And such symmetries abound after we pursue the intensity of unfolding dimensions. In this sense, zero is 'produced by a thought experiment, by a compensation device capable of enveloping a Two with minimal means' (p. 82, *emphasis in original*). Additionally, it can be 'understood as a "middle", as the product of the neutralization of +A by –A, which also makes it possible to open zero out into two branches, and it is just this allusion to opening that permits the conquest of the clichés associated with iteration' (p. 82).[7] This zero is the join of two fingers, the fulcrum of the see-saw. It is not the clichéd taking away of the last cookie, because it requires the carving out of a new space that, once created, generates new mathematical objects (negative cookies!). One runs the risk of taming the mobility of zero by failing to acknowledge its generative power, which is not just in the middle, as in the number line in Figure 2.3a, but is also in the singular fulcrum that originates motion, as in the diagram in Figure 2.3b (taken from Châtelet, 1993/2000, p. 83, based on Argand's work). In the figure on the right, the zero comes first and the other numbers fade from focus. What is intriguing about Argand's diagram is the way that zero evokes new and as yet unscripted directions – new branches of mobility that might invent alternative symmetries of choice.

[7] In referring to 'iteration', Châtelet means the kind of mechanical repeated juxtaposition of numbers that characterizes addition and subtraction (and some conceptions of multiplication).

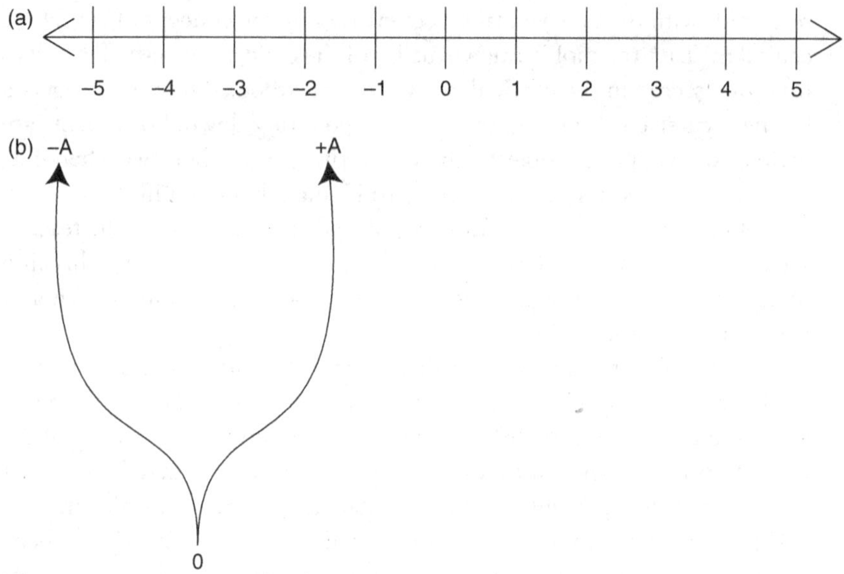

Figure 2.3. (a) The number line; (b) accentuating the role of 0.

The concept of zero that we are proposing resonates to a certain extent with the number line model that many mathematics educators have been advocating (see Saxe et al., 2007), which actualizes the arithmetic-is-motion-along-a-path metaphor identified by Lakoff and Núñez (2000). We see the number line as a powerful diagram for evoking the concept of zero that we have been discussing, especially when used in its potential form (not as the marked segment from 1 to 10 that can be found on the desks and the walls of many primary school classrooms, but as a concept that evokes the infinite direction and density of the mathematical number line).

Drawing on Châtelet, we have offered a few examples of the way we might interpret Barad's relational ontology in the context of mathematics. In the ensuing chapters, we show how this approach to concepts could be used to rethink mathematics curriculum and instruction. We have focused on concepts in part because of their analogical relationship to the concepts of physics that Barad describes. However, we have also focused on concepts because of their singular importance in mathematics and in the teaching and learning of mathematics. Furthermore, in keeping with the notion that concepts emerge from boundary-making practices, we are forced to reckon with the idea that concepts are provisional determinations, which means that they cannot be fixed within the assemblage that involves the mathematician, the learner, the teacher, the diagram, etc.

How mathematical assemblages come to matter

Barad (2011) aims for more than simply including non-humans (like cubes and concepts) as agents of change. She also calls into question our boundary-making practices by which the non-human is excluded from agency, because these undercut any attempt to increase inclusion:

> Rather, the point is that the very practices of differentiating the 'human' from the 'nonhuman,' the 'animate' from the 'inanimate,' and the 'cultural' from the 'natural' produce crucial materializing effects that are unaccounted for by starting an analysis after these boundaries are in place. (p. 124)

It is important to state that the differences between human and non-human are hugely important and are not denied here. But troubling the category of the human does help us begin to examine how those differences and boundaries are sustained and reconfigured through particular kinds of activity. In this book, we are interested in how mathematical activity is implicated in this complex process of becoming human. The analysis in the last section dwelled on the ontological prowess of concepts, showing how one might engage with their material power, in addition to their rule-bound behaviour. But the question as to how this prowess partakes in processes of becoming human remains unanswered. And although this is a huge question that we cannot answer fully – and to which we will return in ensuing chapters – we want to outline here how the kind of analyses in the previous section can be expanded so as to study the mattering of larger (more recognizably human) assemblages. As we said earlier, inclusive materialism taps into phenomenological studies of embodiment, following the micro-visceral movement of mathematical activity, while also attending to the enduring political forces that operate through material assemblages. Our hope is that this dual emphasis on the micro and the macro will furnish a more robust study of the mathematical body. In other chapters, we use policy documents and video data from classrooms to pursue this kind of analysis, showing how affect circulates across and through bodies, but in this chapter our focus is on the mathematical concept. Tracking the links between mathematical concepts and cultural formations is quite a challenge. Historians of mathematics are often tentative to make such arguments, opting for historical accounts of mathematical change without referring to how these changes link to other social changes. Because we work within the field of education, we are not allowed such luxuries. We need to show how political forces flow across mathematical concepts in order to show how assemblages are formed with established or strongly determinate forms of social power. We see our inclusive approach to concepts as complementary to socio-political work, in that we are trying to attend to various kinds of agencies that have been denied value or status in the past. Our efforts to destabilize concepts, as we did in the previous section, and to redistribute agency across a field of potential, feed into our commitment to map political configurations of life.

A pedagogy of the concept thus entails a historical account of the life of the concept. For instance, consider projective geometry's point at infinity, which seems to be a candidate for a highly abstract, immaterial entity, and is one associated with incredible controversy and debate in the history of mathematics. One of the materializing effects of the boundary-making practices involved in mathematics education involves asserting that this point at infinity is an 'as if' entity that 'does not really exist'. However, as we will see in more detail in Chapter 3, there are other boundary-making practices that posit this point as a cut-out of the plane, often marked by a horizon line that is reached by following a trail of dotted lines (see Figure 2.4). Like Bohr's atom, which is indifferent to classical assumptions about space and time, this point at infinity asserts its materiality by living on the plane and receiving the incoming lines of projection. As we will argue in Chapter 3, its materiality on paper (or on a screen) cannot be separated from the gesturing hand that reaches out and sweeps, points, scratches or excavates the horizon line into being. This is the argument that Châtelet offers in his study of the generative nature of diagrams. For the moment, though, what we want to exemplify is Barad's notion of boundary-making practices and how they might be relevant to mathematics. In Chapters 6 and 7, we discuss the political intersections between these mathematical acts of materiality and theories of perception. But here we want to draw attention to how asserting the materiality of the point at infinity is also a political act, forming part of a social assemblage. The effects of this materialization of the point at infinity are wide-ranging, not least of which was, of course, the whole emergence of perspective art.

We raise this example here because it underscores how techniques of observation have changed over time (thus serving our interest in exploring the consequences of quantum ontologies) and how these changes are linked to the development of mathematics. As the art historian and cultural critic Jonathan Crary (1990) suggests, shifts in techniques of observation are inseparable from the 'massive reorganization of knowledge and social practices' that have transformed in myriad ways 'the productive, cognitive, and desiring capacities of the human subject' (p. 3). The point at infinity allows

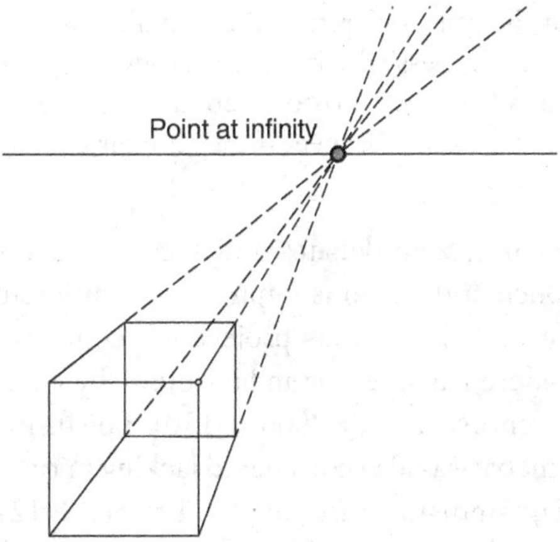

Figure 2.4. The point at infinity in the perspective drawing of a cube.

for a new kind of observing subject who is both 'the historical product and the site of certain practices, techniques, institutions and procedures of subjectification' (p. 5). Crary suggests that we draw on the etymology of the word *observare* – 'to conform one's action, to comply with' – so that we better understand how observing entails 'observing rules, codes, regulations, and practices' (p. 6). The point at infinity is a kind of optical device that transforms the status of the observer, while embedding philosophical, aesthetic and institutional concerns into the mathematics. The point at infinity redefines visibility, thereby reconfiguring the relationship between geometry and the eye and setting the stage for a remapping of the body and its organs. For instance, one might argue that the point at infinity is the site where the relationship between touch and vision becomes newly configured, where a new theory of integrated perception brings forth a virtual dimension on the page. Indeed, Descartes, Berkeley and Diderot's theories of vision were based on analogies with touch, and the sense of touch remained an integral part of classical theories of vision throughout the eighteenth century. If the point at infinity contradicts Euclid's parallel postulate, it does so through a reconfiguring of the body and a simultaneous reconfiguring of the perceptual coordinates of the body in the world. In a reversal of other historical accounts, Ivins (1938/1975) argues that the parallel postulate reflects an ancient Greek valuing of the tactile over the visual:

> The underlying tactile assumptions of Euclidean geometry are excellently exhibited by its basic postulate about parallel lines. If we get our awareness of parallelism through touch, as by running our fingers along a simple

> holding, there is no question of the sensuous return that parallel lines do not meet. If, however, we get our awareness of parallelism through sight, as when we look down a long colonnade, there is no doubt about the sensuous return that parallel lines do converge and will meet if they are far enough extended ... (p. 8)

Whether one agrees with Ivins or not, such debates underscore how the changing relationship between touch and vision is implicated in different geometries. The point at infinity, and the various projective geometries that were enabled by it, were considered 'dangerous and unsound' by many mathematicians in the nineteenth century (Smith, 2006, p. 150). The fusion of touch and eye in projective geometry was also considered lacking in rigor because of its over-reliance on diagrammatic reasoning (de Freitas, 2012). This historical example, given only a brief outline here, indicates how one might begin to unpack particular mathematical practices to see how they entail particular configurations of the body. Crary (1990) argues that the separation of the senses occurs later in the nineteenth century through an industrial remapping of the human body and the development of new optical devices, like cameras. In Chapters 6 and 7, we return to this issue of how the sense organs are differently configured in different time periods, and we discuss how this relates to particular developments in mathematics and mathematics education.

Consider as another example the notion of number and its changing ontology over time. In studying number historically, Badiou (2008) is interested not only in how different number systems have been defined (which was often in response to the discovery of new kinds of numbers, such as negative and irrational numbers), but also in the changing status of three particular numbers: zero, one and infinity. He contrasts two approaches to the theory of number, that of Dedekind and that of Frege; both Dedekind and Frege make different articulations of these three particular numbers, each of which has different materializing effects. According to Badiou, Dedekind's approach involves first positing the whole chain of numbers (the infinite) and then construing the number one as the starting point, the first number in the chain. All other numbers (2, 3, 4, ...) are constructed out of the whole chain, effectively 'cut' at particular places along the chain.[8] This means that the infinite precedes

[8] Some pedagogical texts describe 'intuitive' ways of thinking of real numbers as cuts on the number line. This exemplifies a particular boundary-making practice in which rational numbers – and later irrational numbers – are taken to be immaterial but can be 'represented', for epistemological ease, as geometric locations on a line. A new materialist approach would describe these cuts not as a theory of representation, but as folds that distort and twist and enliven new and sometimes paradoxical potentialities of the continuum.

the finite and that zero is more or less excluded, or at least demoted, from the very meaning of number. Of course, Dedekind does not ignore zero altogether, but in his articulation, zero has a derivative status.[9]

If Dedekind embraces the infinite and demotes zero, Frege offers quite a different ontological approach. He starts with zero as the rock of the whole edifice', while the number one is a mundane follow-up number that plays second fiddle. But then the infinite is the prolongation of number, an act of repeated addition, rather than a force that animates the number line. As Badiou suggests, Dedekind sculpts, taking numbers out of the infinite, whereas Frege paints, adding numbers onto the blank page. In order to appreciate Barad's argument in this context, it is important not to assume that this talk of sculpting and painting is merely metaphorical. These different ways of making number are in fact different material acts whereby the ontology of the number concept is at play.

With this focus on number, we do not simply aim to show how mathematical concepts grow and change historically, which any historical investigation will reveal. Rather, we wish to show that the particular determinations of number at any given time have materializing effects on *learners*, so that they become part of the material-cultural entanglement that we are proposing. In the case of number, Dedekind and Frege are determined to rethink the concept outside of any psychological models, and they desperately want to disassociate number from conceptions of space, because of concerns with the continuum and its potential for paradoxes. Both try to rescue a discrete notion of number, and then, particularly in the case of Frege, build a logical foundation for mathematics based on this discretisation. This very brief discussion reveals how an ontology of number is linked to the emergence of logicism, but also how this new concept of number affords power to computational proofs, as well as set-theoretic ways of organising mathematics curricula in the twentieth century, inaugurating, arguably, the tortuous experience of New Math. A pedagogy of the concept must become historically aware of these cultural links between particular concepts and arger cultural assemblages. In both of the next two chapters, we further develop our use of Châtelet's historical insights to reorient the way we study mathematical activity in classrooms.

[9] Badiou writes that Dedekind 'abhors the void and its mark, and says so quite explicitly: "[W]e intend here for certain reasons wholly to exclude the empty system which contains no elements at all"' (p. 14). Niremberg and Niremberg (2011) criticise Badiou for misinterpreting Dedekind's attitude towards zero, noting that Dedekind goes on to write: '... although for other investigations it may be appropriate to imagine such a system' (p. 590).

3 Diagrams, gestures, movement

In mathematics education research publications, it is increasingly common to find screenshots of student or teacher gestures, along with descriptions of how they communicate certain mathematical ideas – either in concert with speech or, more often, preceding it. Several scholars have worked at describing, categorizing and even prompting gestures in the classroom in order to understand better teachers' and students' non-propositional and body-centric modes of thinking. Gilles Châtelet is also very interested in gestures, but instead of studying the gestures of mathematicians in the classroom, he analyses their diagrams. Thus, Châtelet refuses to separate gestures and diagrams; it is through their interplay that he shows how 'immovable mathematics' can come to be seen as a deeply material enterprise. In Châtelet's gesture/diagram investigations, we find compelling evidence of the way the body partakes of mathematics, as well as a mathematically grounded instantiation of many of Karen Barad's ideas.

Consider the diagram shown in Figure 3.1, which was produced by a geometry student who was asked to make a drawing that shows the movement of a circle. As we will show in this chapter, Châtelet might have examined this diagram for how the diagrammer was entangled with the mathematics, and for how the surface of the paper captured mobility and engendered dimensions. Châtelet might ask: What work is the dotted line doing in this diagram? Is it the trace of a gestured arc, moving away from the horizon? Châtelet would be far less interested in reading the diagram only as a representation of mathematical objects (in this instance, possibly, two sets of concentric arcs), nor simply as an aid for solving a given problem. His overarching goal is to study inventive moments in mathematics by examining the gesture/diagram interplay from which they emerged.

We begin the chapter by discussing the particular ways in which Châtelet described gestures, diagrams and their interplay, as his approach

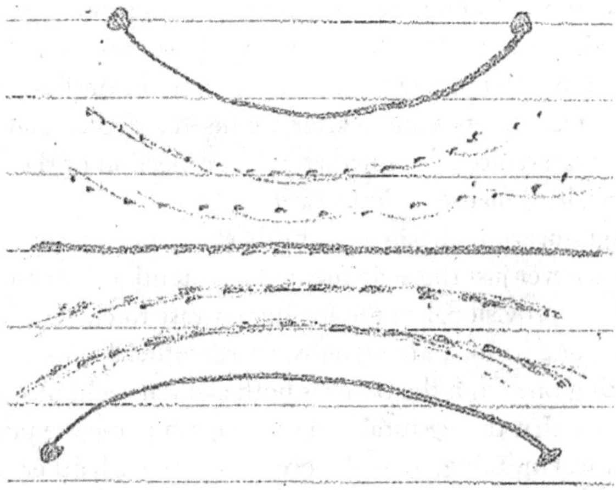

Figure 3.1. Diagram with dotted and solid arcs.

differs quite significantly from the way gestures and diagrams are studied in mathematics education research. We then provide a few examples of the inventive moments Châtelet studies in order to exemplify his methodological approach, and we complement these with additional diagrams that underscore the ontological implications of his materialist assumptions. Finally, we consider diagrams produced in a classroom context (like the one from which Figure 3.1 was taken) in order to explore how Châtelet's methodology can be applied in a school mathematics context. In our analyses of the diagrams produced by students, we develop several implications of inclusive materialism for school mathematics.

On gestures and diagrams

Châtelet selects certain episodes in the history of mathematics and physics to show how particular diagrams – what he terms 'cutting-out gestures' – have erupted during inventive thought experiments to reveal both mathematical agency and ontology. In other words, he uses these historical episodes to explore ontological questions about the relationship between the mathematical and the physical, as well as psychological questions about what it means to do mathematics. He argues that the study of such gestures helps us undo some of the troubling consequences of the Aristotelian division between movable matter and immovable mathematics.[1] The fear and loathing expressed by Russell towards the very idea of the motion of a point in space is an obvious expression of this tradition. For Châtelet (1993/2000), the attempt to separate immovable mathematics from movable matter is 'a rational account of illusion' (p. 14). In other words, mathematics and matter are fused together, and any attempt to deny the materiality of mathematics is a reflection of the desire for an ideal and unsullied mathematical world.

The diagram, argues Châtelet, is by its very nature never complete, and the gesture is never just the enactment of an intention. The two participate in each other's provisional ontology. In contrast to current work on gestures, on the one hand, and diagrams, on the other hand, Châtelet insists that separating one from the other is both awkward and possibly misleading. He argues that the gestural and the diagrammatic are pivotal sources of mathematical meaning, mutually presupposing each other and sharing a similar mobility and potentiality (we will show in the final three sections of this chapter how a diagram such as Figure 3.1 can be thought of as having mobility and potentiality). In other words, gestures give rise to the very possibility of diagramming, and diagrams give rise to new possibilities for gesturing. Châtelet claims:

> A diagram can transfix a gesture, bring it to rest, long before it curls up into a sign, which is why modern geometers and cosmologers like diagrams with their peremptory power of evocation. They capture gestures mid-flight; for those capable of attention, they are the moments where being is glimpsed smiling. (p. 10)

For Châtelet, diagrams 'lock', or 'capture', gestures. 'Capturing' is contrasted to 'representing' in that the latter is bound to a regime of signification that curtails our thinking about diagramming and gesturing *as events*.

> Like the metaphor, they [diagrams] leap out in order to create spaces and reduce gaps: they blossom with dotted lines in order to engulf images that were previously figured in thick lines. But unlike the metaphor, the diagram is never exhausted: if it immobilizes a gesture in order to set down an operation, it does so by sketching a gesture that then cuts out another. (p. 10)

In linking diagrams to metaphors, Châtelet points to their evocativeness, but he also insists that diagrams are 'a little less impertinent' than metaphors, because 'it is always possible to seek solace in the mundane plotting

[1] We discuss this troubling divide in more detail in Chapter 8, where we focus on Châtelet's important notion of the virtual.

of their thick lines', and they are more faithful, because 'they can prolong themselves into an operation which keeps them from becoming worn out' (p. 10). Diagrams are more than depictions or pictures or metaphors, more than representations of existing knowledge; they are kinematic capturing devices, mechanisms for direct sampling that cut up space and allude to new dimensions and new structures. By adding a dotted line to the paper, a new dimension can be brought into being; for example, an arrow might forge out new temporal relationships between objects. These excavations enable the virtual and the actual to become coupled anew.

The power of the gesture is in the unanticipated accuracy of its 'strike'; the gesture is never entirely captured and there is no algorithm for determining it. There is no rule that enunciates and decomposes the act into a set of repeatable moves[2]; a gesture is allusive and allegorical, and it inaugurates 'dynasties of problems' (p. 9). A gesture is more than simply an intention translated into spatial displacement, for there is a sense that 'one is infused with the gesture before knowing it' (p. 10). The inventive gesture is outside the domain of signs and signification, insofar as signs are coded and call forth an 'interpretive apparatus' that exists prior to them (Rotman, 2008, p. 36). This way of seeing the gesture enables us to comprehend the body as more than a 'silent, dumb, a-rational and emotional' object (p. 48). It is precisely the cracking open of this silence that allows us to debunk the mystical interiority presupposed by the Kantian valuating of the verbal. The body and its silence are no longer governed by the linguistic and the sayable. Rotman is careful, however, to declare that gesture will always exceed textuality, signification and the hermeneutics of decipherment. The embodied gesture will always exceed attempts to reduce it to a science of gesturology. If a gesture functions in terms of reference or denotation or exemplification, it is already stale and domesticated. Châtelet is concerned with gesture as a kind of interference or intervention that has driven mathematics and the sciences forward, not as a sign divorced from the mathematical event, but as a dynamic process of excavation that conjures the sensible in sensible matter.[3]

[2] Châtelet's interest in gesture differs in some ways from that of McNeill (2003). In fact, Châtelet is less interested in any sort of classification or complete description of gesture than in the implications of the gesture as inventive. The gesture is assumed to be an intermediary from body to diagram. Châtelet keeps a respectful distance from any kind of propositional, classificatory attempt at describing it, partly because of his insistence on the gesture as allusive, evocative and even covert.

[3] Diagrams can also include notation and, as we will discuss in Chapter 8, Châtelet is also interested in the way notation can be used to interfere with linear readings of tasks and to conjure new relations and metaphors.

Thus, the concept itself is entailed in the hands that gesture, the mouth that speaks, and the affect that circulates across an interaction. This complex assemblage of partial agents and provisional organs emerges in a process of becoming, an indefinite process of individuation enmeshed with speech codes that leverages ambiguity and somatic force. Our emphasis on the mobility of mathematical concepts helps us see how mathematics often emerges through this entanglement. The making of a circle, for instance, is a simple gesture that particular assemblages and bodies are inclined to produce without being told or instructed, and frequently without any prescribed verbal equivalent or operating program. For many researchers studying mathematical activity, however, a gesture is not yet fully mathematical in that it needs to be 'mathematized' into formal mathematical terms. The supposed imperfection of the gesture, however, needs to be problematized, because it insists on a lack (of axiomatic grounding) rather than seeing the gesture as a surplus (of potentiality). Regimes of signification (Euclidean, Cartesian, topological) come to saturate this generative gesture, affixing it to a frame of reference. Archimedes will bemoan the disturbing of the diagrams of his perfect circles as a soldier ends his life during the sacking of Syracuse (Doxiadis & Mazur, 2011). Yet somehow the circular gesture outlives all of these trappings and may in fact be the aggravating rub that troubles their complacency, refusing to be squared with 'common sense'.

Perhaps other mathematical concepts can be studied for how they maintain their gestural indeterminacy, similar to the way in which Husserl imagined a 'proto-geometry', but without his commitment to the disembodied ideal. Merleau-Ponty (1962/2004) comes close to this approach in describing his own engagement with a diagram of a triangle, pointing to the indeterminacy and potentiality that is induced by the act of diagramming: 'my perception of the triangle was not, so to speak, fixed and dead, for the drawing of the triangle on the paper was merely its outer covering; it was traversed by lines of force, and everywhere in it new directions not traced out yet possible came to light. In so far as the triangle was implicated in my hold on the world, it was bursting with indefinite possibilities of which the construction actually drawn was merely one' (p. 185). The point here is that mathematical concepts can be studied for their ontogenetic force, rather than as elements of a system of representation.

Like the gesture, the diagram is a kind of potential and is never entirely actualized, standing somehow on the outside of signification, transfixing as it does the gesture. 'The diagram invites an erasure, a redrawing, a "refiguring"' (Knoespel, 2000, p. xvi). Every diagram may be reactivated through our engagement. It is important to note that diagrams are at the heart of

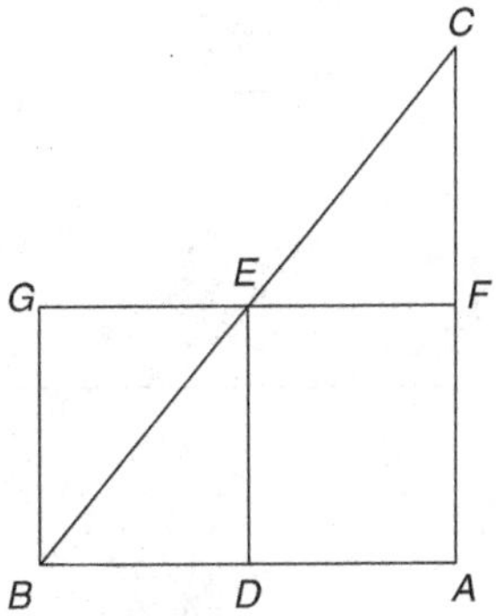

Figure 3.2. Oresme's configurationes for linear qualities unites extensive (time on the horizontal) and intensive (speed on the vertical) quantities so that distance can be calculated in terms of area. The area of triangle ABC gives the length travelled in time between B and A (equal to the area of BAFG).

Châtelet's historical study of the emergence of new mathematical ideas, for it is the diagrams, and not the gestures, that have survived. For instance, he shows how the fourteenth-century kinemathematician Nicole Oresme revolutionized the study of the motion of motion by generating new diagramming techniques. Oresme referred to these diagrams as 'configurationes', by means of which he was able to study the spatial and diagrammatic rendering of various physical and mathematical concepts (Clagett, 1968). The most historically significant of these configurationes (Figure 3.2) were those that used the geometry of similar figures and their ratios to show the equality of a right triangle (which represents uniform acceleration) with a rectangle (which represents uniform motion), constructed and superimposed as the velocity of the middle instant of acceleration.

Rather than conceptualizing diagrams as idealizations of mathematical relationships, however, Châtelet invites us to see diagrams as invoking a dynamic process of excavation that conjures the virtual in sensible matter – in Oresme's case, the virtual acceleration of an object is conjured through an area diagram. In other words, the inventive diagram is an action that literally breaks down previously taken-for-granted determinations of what is sensible or intelligible and actually carves up matter in new, unscripted ways. According to this approach, the diagram is a physico-mathematical entity, with elasticity and mobility, that can 'cut out' new dimensions in the plane – 'the plane is made flesh, as it were' (Châtelet, 1993/2000, p. 34). It is physico-mathematical in the sense that it eludes the usual material versus mathematics distinction of Aristotle by uniting in one entity the actual and the virtual – the diagram and the gesture that produced it.

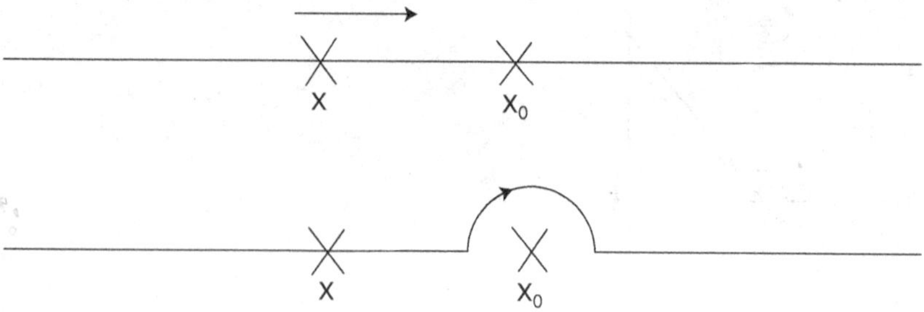

Figure 3.3. Châtelet's 'cut-out' diagram.

Châtelet also investigates the works of Cauchy and Poisson on singular points. The diagram in Figure 3.3, which concerns $f(x) = \frac{1}{x - x_0}$, might easily be read as a *representation* of a singular point in a complex plane – designating the point as merely two-dimensional. This diagram, however, was generated by Cauchy in the nineteenth century as part of his residue method for integrating around singular points. The little circle around the point departs from the real plane and cuts into the complex plane, excavating and giving density to the point as though it were a potential hole in sensible matter. For Châtelet, the diagram constitutes the point as a material bump in the surface of the page. Taking a new materialist point of view, Châtelet reads x_0 as being made flesh by a 'cut-out' in the complex plane around which the point is now enveloped. Such a diagram, for Châtelet, can only grow out of an accompanying gesture – not a tentative deictic pointing at something on the surface, but an actual, physical creasing or cutting out which marks up the surface and conjures its virtual folds. Such an interpretation of how this diagram works allows us to grasp its historical significance in complex analysis better, and it points to the ontological implications of such inventive diagrams. Indeed, as part of his investment in the study of diagrams as 'physico-mathematical beings', Châtelet imagines a world in which the point is a sensible point, a point set ablaze by motion and depth.

Mobile Mathematics

The *virtual* plays a central role in this new approach to gesture and diagram, because it marks that which is latent or ready in a body. In the case of the diagram, the virtual *is* the mobility that is presupposed by an apparently static figure, a generalised mobility that was central to its creation in the first place. In other words, the virtual or potentiality of a diagram consists

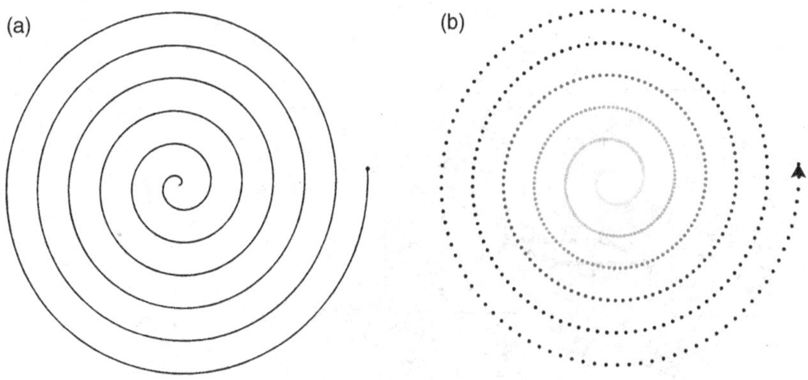

Figure 3.4. Archimedes' spiral (a) the static form; (b) a more dynamic representation.

of all the gestures and future alterations that are in some fashion 'contained' in it.[4] Consider, for instance, Archimedes' spiral – a curve generated by tracing a point as it moves away from a fixed point at a constant velocity along a straight line, which is rotating around the fixed point at a constant velocity. Figure 3.4a depicts the static version of the diagram, as it is shown in most textbooks. In Figure 3.4b, the path travelled by the point can be seen in the faded traces, giving the spiral a more temporal, dynamic feel.

The diagram on the left (Figure 3.4a) contains all the motion and gesture that was entailed in its construction, and yet we perceive only the static image. The virtual is 'still' there and can break out of the static diagram if properly cut open. Indeed, abstractions cannot be divorced from the sensible matter of the spiral. The diagram captures the spiral by carving up a certain space (the winding line), while embedding it in space (the plane of the paper or computer screen). It is not a representation of a spiral, but rather a device that traps and contracts the material world.

Consider also the following visual proof in Figure 3.5 (a proof that line segments joining the adjacent centres of squares built on the sides of a parallelogram will always form a square), which seems to convey a greater sense of motion. This diagram consists of at least three perceptual layers: a virtual layer conjured through the dotted line that elicits the mathematical relationship; an actual layer that presents the shaded figures; and a third virtual layer that conjures the act of tilting or hinging because of the repetition of oblique and acute angles.

[4] In Chapter 8, we develop in detail Châtelet's notion of the virtual, which has strong links to the potential. As we will show, any virtuality will emerge out of this potential.

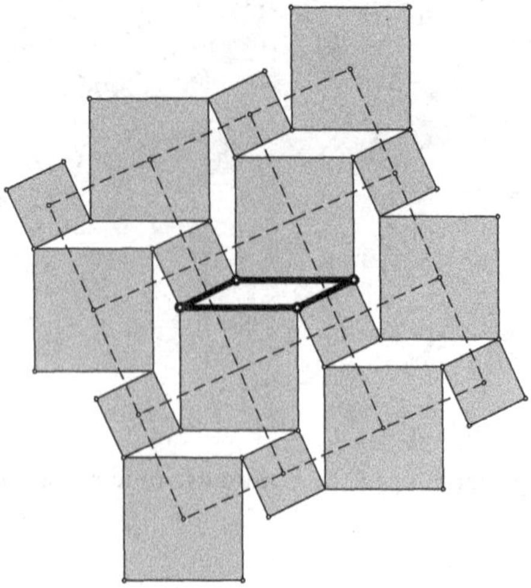

Figure 3.5: Visual proof.

Again, this diagram is not, according to Châtelet, a representation of a proof, or at least not *only* a representation of a proof. Reducing a diagram to a representation robs it of its materiality. As Rotman (2008) writes:

> [It] ignores the corporeality, the physical materiality (semiotic and performative), as well as the contemplative/intuitive poles of mathematical activity; and in so doing dismisses diagrams as mere psychological props, providing perceptual help but contributing nothing of substance to mathematical content. (p. 37)[5]

One can see in Châtelet's approach an attempt to rethink matter itself as well as the relationship between the virtual and the actual. Deleuze (1993) argues that this approach to metaphysics (in the sense of what might exist beyond the actual, or the physical) is best explored through the study of particular areas of mathematics that have forced us to reconceptualize the relationship between the virtual and the actual, pointing to the work of Galois, Riemann and others, in areas such as algebraic topology, functional analysis and differential geometry. Both Châtelet and Deleuze argue that Leibniz (and 'Baroque mathematics') offers an alternative starting point for rethinking the relationship between immovable mathematics and movable

[5] Although he is coming from a very different philosophical point of view, Netz (1999) is also at pains to point to the more-than-psychological role of the diagram in ancient Greek mathematics. In his more recent work (2009), he tentatively suggests that those diagrams were performed by ancient Greek mathematicians, thereby breathing mobility into these immanent icons that were long-assumed to be static.

matter. For Leibniz, motion is constitutive of bodies, and point of view and perspective, rather than extension, are definitive of substance. Leibniz sees the world as comprised of an infinite number of monads, each with its distinct point of view and each 'compossible', or presupposed by all the others. The ontology of monads feeds into Leibniz's theory of a relative space–time continuum (or spatium) that is conceived as a fluid of relations and differentials (Parkinson & Morris, 1973). The monadology is a metaphysical counter to Descartes' passive nature and Newton's erasure of space and time through absoluteness.

Within this fluid world of differential relations, actions of any kind are conceived as folds in the spatium. The cutting-out gesture creates a new fold on the surface; it pleats and creases matter, and it generates depth and even interiority (Deleuze, 1993). Cutting-out gestures interfere with the surfaces that determine the relationships between the material and the virtual. Both gesture and diagram, according to Châtelet, are akin to a thought experiment, which 'separates and links, and therefore becomes an articulation between an exterior and interior' (p. 32). The dotted line of the diagram intimates or suggests the making of a new inside/outside – the folding of space into new surfaces. Although Châtelet calls these newly made surfaces 'cut-outs', their individuation is never apart from the spatium; the cut-out simply folds, creases and partitions matter and mind in such a way that the unthought is able to enter onto the page.

Consider, for instance, the circle and the trefoil knot in Figure 3.6. The visual breaks or overlaps in the knot conjure an effect of layering where Cartesian plane geometry would have imposed an intersection. Topological diagramming forces us to decode the overlapping of the knot, which would normally be a three-dimensional act, in terms of a virtual dimension within

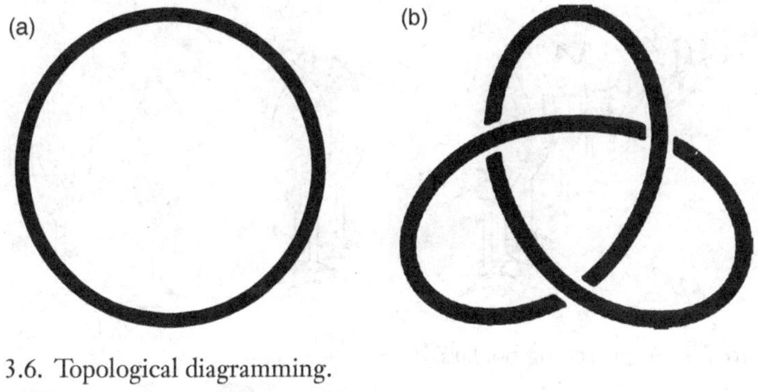

Figure 3.6. Topological diagramming.

the two-dimensional plane, as though the plane were suddenly able to accommodate a new kind of depth.

Châtelet (1993/2000) notes that scientists reflecting on knot diagramming in the nineteenth century already knew that these diagrams were not 'simple illustrations' and that they pointed to the event-structure of intersection and would indeed 'smash the classical relationship between letter and image' (p. 184).

> [G]eometric beings are not what remains when all individuation is ignored, instead they must be recognized as part of more ample physico-mathematical beings, which force us to reconsider the relationship between logical implication and real implication. (Châtelet, 1993/2000, p. 32)

In all the examples of diagrams discussed thus far, there is a *hinge-horizon*, where the surface seems to end. To decide upon a horizon is to determine a metric that codes the space, a metric that then domesticates the mobility of bodies. A hinge-horizon is also a line marked on a surface whereby the viewer might begin to grasp the infinite in the finite. For instance, the vanishing point in a painting constructs a hinge-horizon and makes the infinity of space perceptible (see Figure 3.7). In such paintings, the depth of space is conjured through a knitting together of vertical and horizontal oblique lines. 'With the horizon, the infinite at last finds a coupling place with the finite' (Châtelet, 1993/2000, p. 50). As an example, consider how perspectival drawing joins the infinite and the finite together in a continuum of similar figures.

What do we see when we look at this diagram? Do we see repeated similar but dilated figures, or do we see one figure moving towards or away from us? Can we train our eyes to see both? Châtelet suggests that we try to perceive the motion that binds the figure at the forefront to the faded but similar figure found in the virtual dimensions behind the page. Indeed, it is

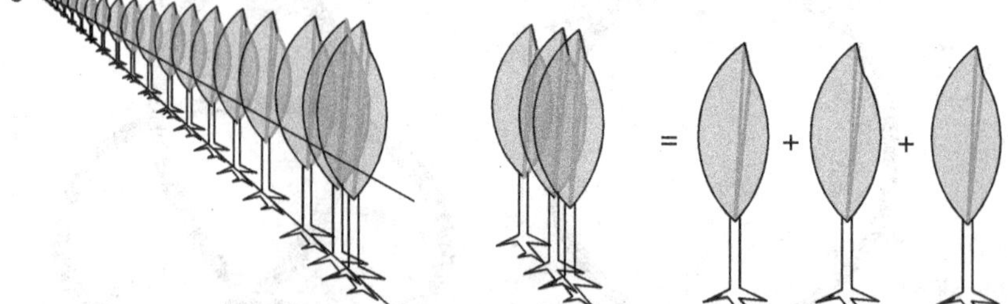

Figure 3.7. Approaching poplars.

as if the figure were constituted by this movement *of* movement – a form of acceleration, of expanding iteration – whereby the figure comes out of depth and into proximity. In this fluid world of differential relations, extension is garnered through motion; that is to say, length becomes intelligible by way of a moving body along a vanishing line. As Châtelet declaims, 'No length without velocity!' (p. 49). Nothing, therefore, *inheres* in the horizon – figures come into place through the mobility that relates one to another. Motion is primary or constitutive – and the horizon is an allusion.

There is much, however, which *adheres* to the horizon. Like any other virtuality, once it has been actualized, it remains:

> Once it has been decided, one always carries one's horizon away with one. This is the exasperating side of the horizon: corrosive like the visible, tenacious like a smell, compromising like touch, it does not dress things up with appearances, but impregnates everything that we are resolved to grasp. (Châtelet, 1993/2000, p. 54)

Despite its compromising aspect, the horizon is an elastic hinge-horizon, inviting dilations and compressions, folds and distortions. In articulating a horizon, one instantly perceives its enveloping character and must begin the work of problematizing it as stasis. Indeed, citing de Broglie and Einstein, Châtelet shows how even the concept of a body at rest has been made problematic through relativity theory, wave theory and the defining of mass in terms of angular momentum.

The stasis and confining aspect of the hinge-horizon is undone by way of Châtelet's 'diagrammatic experiments' (p. 63). But how does one develop a set of devices for folding surfaces or for creating points of inflection and singularity that resist the closure of the enveloping eye? How might we invite the radical gestures of invention – the hand that strikes so accurately in some unprescribed manner – under the watchful, definitive eye that longs for its horizon? How can the hand break out from under the vigilant eye? In the next section, we describe an experiment in which we attempted to set the stage for inventive diagramming by students, and we examine – in the manner of Châtelet – what we can learn about students' gestures by reading the diagrams they produced during the experiment.

A diagramming experiment

We hypothesize that one way of developing student diagramming habits includes working systematically with dynamic imagery in order to increase (and perhaps rekindle) the material mobility on which Châtelet's mathematicians drew. In particular, while Châtelet emphasizes the power of gestures to mobilize mathematicians' diagrammatic breakthroughs, he also insists on the diagram's capacity to generate new gestures, or new forms of (imagined) bodily mobility. But unlike the diagrams that Châtelet studies, which are more like sketches and scribbles than finished, iconic symbols, the diagrams of textbook pages tend to drop the idiosyncratic drawing grammar that permits evocative temporal representations. How might such diagrams – devoid of the arrows, dotted lines and cut-outs of Châtelet's examples – generate new gestures, or new mobilities? Dynamic diagrams, on the other hand, like those generated with digital technologies and rooted in time without necessarily using temporal diagrammatic devices, may provide the learner with the desired generative quality.

In an experiment conducted with 28 undergraduate students enrolled in a geometry course intended to fulfil 'breadth' requirements for non-mathematics majors, we borrowed Tahta's (1980) technique of working with Nicolet films. These stop-action films, which were originally created in the mid-twentieth century by a school mathematics teacher, show various geometric objects in motion on a black screen, with no accompanying sound or words (nor hands having drawn the various stills). We chose to work with the clip entitled 'Families of circles in the plane', because that particular three-hour lesson was focused on various properties and uses of the circle. In the clip, a circle appears on a black background, moving around and changing both location and size. A point then appears on the circle, which continues to move while remaining attached to the fixed point. Then, a second point appears (see Figure 3.8a), and the circle becomes progressively bigger (see Figures 3.8b and 3.8c), as if its centre (which is not visible) was moving further and further towards the lower left of the screen along the perpendicular bisector of the two fixed points. Finally, a line (Figure 3.8d) appears; the motion stops for a brief pause, and then an arc appears, which still passes through the two fixed points, and gets progressively smaller (Figure 3.8e).

We chose this film because it seemed to evoke ideas related to projective geometry, namely the notion of a point at infinity. (In other words, if

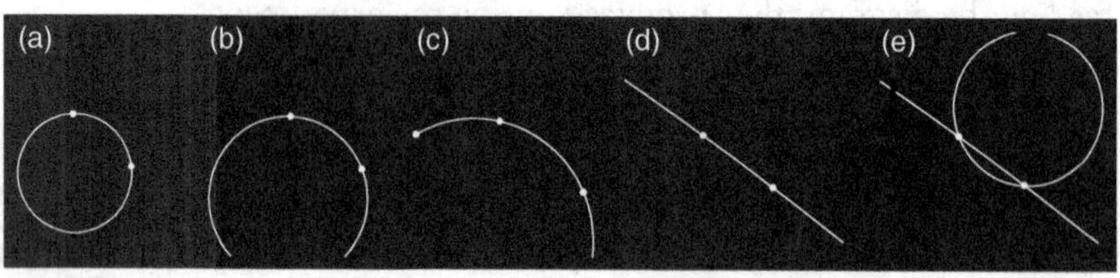

Figure 3.8. Snapshots of the Nicolet film on circles.

the line is seen as a continuous transformation of the circle, then the centre of the circle must be infinitely far away, much like how the vanishing point of a perspective drawing is construed.) Given that the dynamic diagram can be seen as inferring the notion of a point at infinity (or, at least, the idea that a circle can somehow reverse curvature), we offered the diagramming task as a way for the students to explore a virtual and geometrically unfamiliar idea. The film also evoked connections to both of Châtelet's gestural interests in hinge-horizons (the horizon of the point at infinity) and rotation: if the diagram is taken as a head-on view of a three-dimensional situation, the circles can be seen as rotating around the line connecting the two fixed points, so that the line is the visible portion of the circle seen from a perspective that is perpendicular to the plane of the circle. In addition, the film provides a dynamically transforming circle, thus offering an opportunity for the students to think of the circle not just as a familiar shape (as is common in school where one learns how to measure it and to identify its radius, diameter and circumference), but also as a possibly mobile object with certain spatial and temporal behaviour. Finally, we hoped that the film would invite the students to engage with an animated mathematical environment.

In the experiment, the instructor (Sinclair) invites students to watch the clip three times in a classroom setting, and the students are then asked to describe orally what they had seen each time the film concludes.[6] Most students resist seeing a line at the point where the convexity of the circle changes (Figure 3.8d). Several students imagine a three-dimensional configuration, as described in the previous paragraph. When prompted, they do not seem to be concerned with the perspectival problem that such an interpretation produces: If the circle is rotating away, should not it appear elliptical? The following week, the students are asked to make diagrams of the situation with the following prompt: 'Show with diagrams how the circle moves from being concave up to concave down.' More specifically, they are asked to show what happens to the circle, as it moves from its initial position (as in Figure 3.8a) to its eventual position in Figure 3.8e. This is the third class of the semester, and the students had already engaged in diagramming activities in the first two classes, so the prompt is not an unusual one. We note that the diagramming task is offered as an end in itself, and not, as is frequently reported in research, as a means of solving a problem (Nunokawa, 2004; 2006).

[6] Another interesting experiment would be to have students diagram without first making oral contributions, given that speech and listening introduce new modalities.

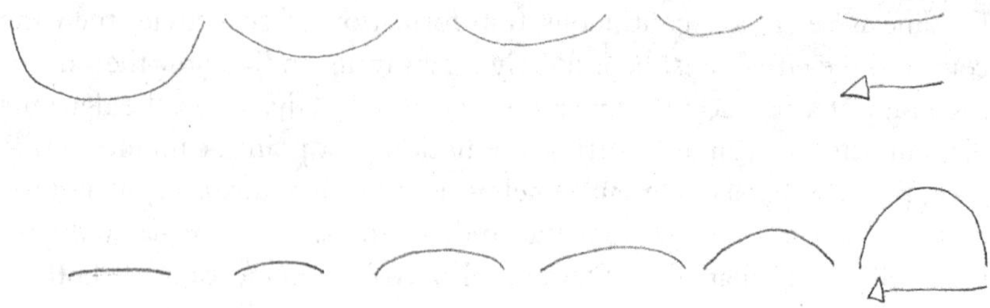

Figure 3.9. A successive framing approach.

We offer here five examples of the students' diagrams. These represent the diverse range of diagramming possibilities that are used. We can see in some of these diagrams precisely what Châtelet finds in significant historical developments in mathematics: inventive 'cutting-out' and dotted-line gestures that interfere with and trouble assumed spatial principles. We analyse these diagrams for evidence of how the student-circle assemblage emerges through distinctive diagramming strategies, as though the students were 'spilling out of themselves' while forming new folds upon the working surface.

In the first example (Figure 3.9), the student uses a successive framing approach to diagramming the circle's changes – one that is similar, in fact, to the stop-action technology of the films. The arrows are used to indicate the direction of time, so that in the first row, the circles are seen as getting flatter until they eventually reach a straight line. The second row shows the circles getting less flat, but again, they are depicted with an ever-moving horizontal line. The diagrams do not clearly show that the series of shapes consist of circles; instead, they seem to focus on the flattening curvature that is approaching the extensive dimension of the line. In Châtelet's terms, the motion of the diagram is along a linear horizon; it neither extends into three-dimensional space (with a fold) nor cuts into the virtual space. However, the horizontal segments shown at the end of the first row and the beginning of the second row are shown as pivotal horizons where things end and also begin. The arrows imply horizontal movement, even though the circle as seen on the screen had a much more vertical motion. One senses the material effect of the lined paper, which is normally used in writing text in the same left-to-right convention that the diagram follows. In this sense, the horizontal line can be seen as a resting spot – emphasized by its maker – between two phases of continual change. But it can also be seen as a break – a break that asserts the impossibility of seeing the changing concavity as a continuous process. As the drawer's hand moves from the left side of the page to the right, it makes a break, announcing a new beginning. We see here in this

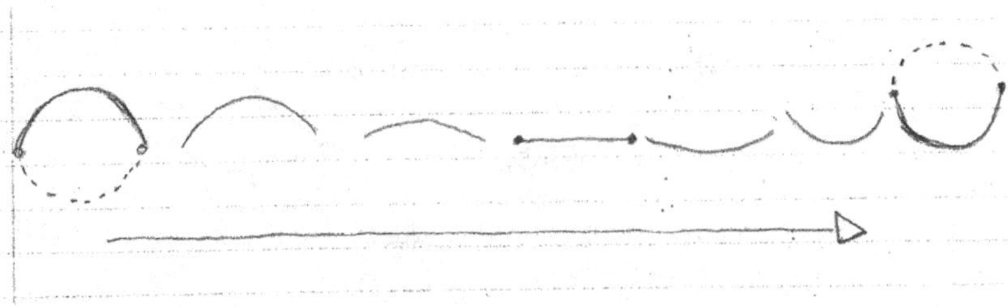

Figure 3.10. Successive framing without discontinuity.

first student example a minimal engagement with the mobility of the circle, or perhaps a discretization of that mobility, much like a stop-action film.

In the diagram shown in Figure 3.10, the student draws on the same successive framing strategy of representing the change in time as a series of discrete shapes. Instead of unfolding over two rows, the transformations occur along the same row, beginning with the half-circle concave down getting progressively flatter, and then turning into a straight line, and finally transforming into a half-circle concave up. Points are used here, as in the film, to indicate positions on the circle that remain fixed, at least throughout the first half of the transformation (floating upward during the second half). These elements were entirely lacking in Figure 3.8. As in the previous example, however, only the arcs are visible. Only the first and the last semi-circles include dotted lines that complete the circle – dotted lines that, in Châtelet's diagrammatic grammar, can serve to couple anew the virtual and the actual. The dotted semi-circles combine with the other elements to construe a slightly more complex assemblage of gesture-diagram. The dotted curves intimate or conjure a future action and thereby draw the hand of the viewer into the diagramming space. That they are only used for the first and last circles, which points to the delicate shift between the student's sense of the real and the virtual. Unlike the solid curves, the dotted curves demand a more embodied reading. They are not to be dismissed as merely subjective or ephemeral, but are rather material traces of the potential aspects of the diagram, and they thus suggest a somewhat enhanced form of embodiment, in that the surface is taken up and cut or folded in ways that disrupt its taken-for-granted status (de Freitas, 2010).

In both of these examples, the idea of the straight line is very strong, and the temporal, mobile relationship is represented as a discrete sequence of steps: The motion is implied by moving towards the right, as if reading a printed page. In comparison, the third example (Figure 3.11) exhibits a strong diagramming power, in that it breaks through the temporal

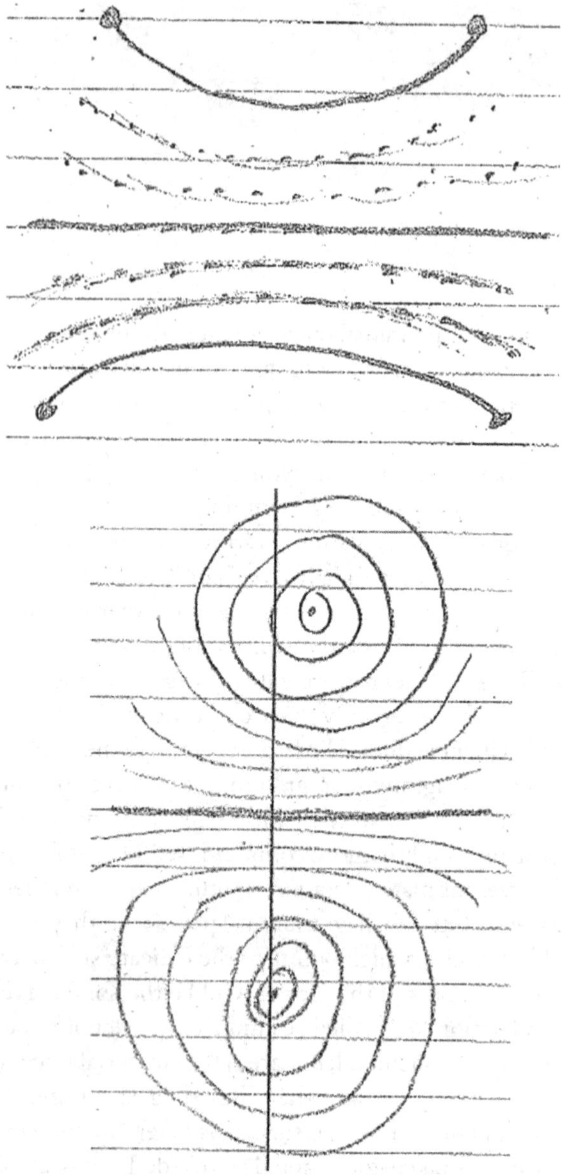

Figure 3.11. A diagram that transcends the temporal constraints.

representational quality of the first two examples. The student's work contains two diagrams – and hence, a kind of diagrammatic study of the situation, rather than a faithful replica of the moving circle on the screen – drawn on a single page. In the diagram above, all the arcs of the circle are shown at the same time, with dashed lines used for the arcs that are getting close to a straight line. The superimposition of the temporal sequence reconfigures the nature of the motion. Here the solid lines indicate starting and ending positions, while the dotted ones are the parts of the circle in motion. The horizontal line is again solid, which might point to the perceived realness of that transition horizon between oppositely curving lines. Indeed, in the film, the circle grows at a constant rate, but the motion pauses when the line appears. For this student, the line and the two circles at the extremes have an actuality, whereas the motion in between is virtual.

In moving to this diagramming strategy, it is interesting that the two fixed points that were visible in Figure 3.10 are now gone. Here the circles are peeling off the line either from the top or from the bottom, as if the arcs are all part of concentric circles, as is more evident in the bottom diagram in Figure 3.11, where the centre of the circle remains fixed while the size changes. However, in the film, the circles were not concentric and the centre was moving. The difference between the student's diagram and the film can be seen in comparing the two diagrams in Figure 3.12.

Returning to the student's diagram, in the bottom diagram of figure 3.11, the dotted lines disappear, and the whole circles become visible. There is no longer a need to distinguish the real from the virtual. In both cases, the concentricity of the circles gives the viewer a sense of depth and suggests that the circles are receding into the background. By drawing concentric circles, the student can keep all the circles on the same side of the horizontal line, thus preserving that line as a cross-over point, a kind of hinge-horizon. Thus, despite the symmetry uniting top and bottom, a barrier remains.

If the top diagram was completed first, perhaps as an exploration of the virtual motion, while the second one, now with the motion

Figure 3.12. Two different ways of imagining circle growth.

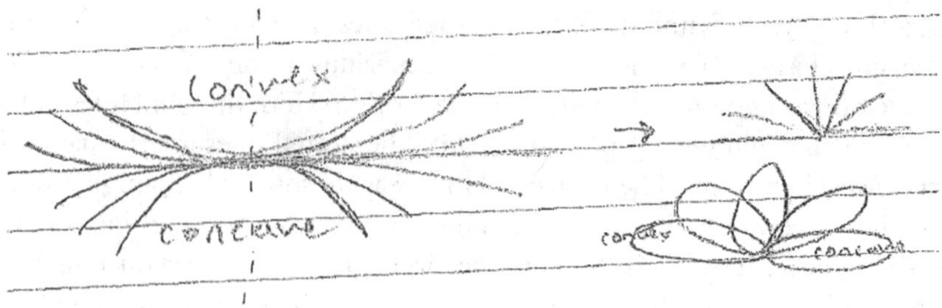

Figure 3.13. Conjuring the path of the centres.

actualized, attempts to capture the fuller spatial situation of circles turning into arcs, and then into a line, and then into arcs and finally into circles again.

The fourth example (Figure 3.13) also consists of multiple diagrams. The first one on the left uses the same strategy as that found in figure 3.11 but the circles are all tangent to each other (and coincident with the horizontal line), with no fixed points. The vertical dotted line conjures the (virtual) line along which the centres of all the circles travel as they get progressively smaller or bigger. Here, the dotted line is used as a diagramming strategy to introduce a new dimension of interest, in addition to the horizontal one. With the top diagram on the right, which seems to show the side-view of a three-dimensional interpretation of the film, the circles are shown as lines that begin flat on the plane perpendicular to the page and rotate around a full 180°. In the two diagrams on the right, the size of the circle is not changing. And the line is presumably the instance when the circle is precisely at a 90° angle to the perpendicular plane. The diagrams on the right thus offer a very different interpretation from the two-dimensional version, in which the circles are getting bigger, while the centre is moving further and further away. In fact, the diagrams on the right convey a certain point of view for the observer (the student drawing) as being beside the circle, as if she were watching a CD case being flipped open. The diagramming studies move from a view of the *xy* plane to a view of the *xz* plane, and finally to a view of the *xyz* space, each transition requiring a perceptual shift. Indeed, each transition, but especially the final, third-perspective view, produces an every growing assemblage, more dispersed than in the third example (Figure 3.11). In all the diagram studies of this example, the idea of fixed points is not apparent, as the asymptotic line takes on primary importance. These previous two examples begin to construe a more provisional gesture-diagram assemblage, altering the very nature of embodiment, and entering into a 'world of pervasive proximity' through shifting perspectives and cut-out

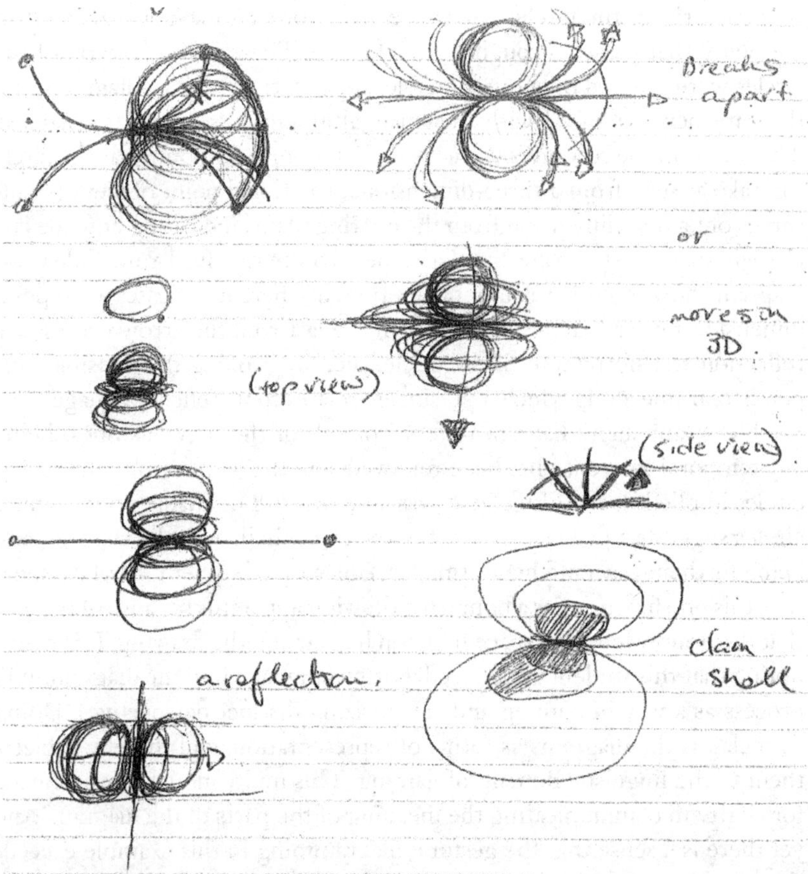

Figure 3.14. Exploding temporality and perspective.

dimensions. This is an immersive subject that is 'increasingly defined by the networks threading through it' (Rotman, 2008, p. 8).

The fifth example (Figure 3.14) has many elements in common with the fourth. However, in addition to offering a more systematic diagrammatic study, it also uses the arrow as a means to achieve new diagrammatic power. Arrows were used in the previous examples, but more as a mode of depicting order (direction) or implication. Here, the arrow is used to evoke new temporal and spatial dimensions. In the top-right diagram, which looks like an octopus,[7] the arrows are placed at the ends of the arcs, gesturing towards the parts of the circle that exist but cannot be seen; the words 'Breaks apart'

[7] We read this diagram as consisting of four rows and two columns, for ease of discussion.

suggest a rip in the circle that is needed in order to achieve the shift in concavity that passes through a straight line. These invisible parts of the circles were granted only questionable status in the previous diagrams, but they are here endowed with existence, although it is a virtual existence. The arrow in the second row shows the direction of motion that the circle can take as seen from a three-dimensional rotational point of view. It thus carves out a new dimension from the existing plane, indicating how the circles will turn into the page. Similarly, the arrow in the third-row, 'Side view' diagram shows a similar rotational motion, but here it indicates a temporal dimension rather than a spatial one. In the last row, the arrow expresses a reflection transformation of the circle, thereby evoking the invisible perpendicular line along which a pre-image relates to its reflected image.

The 'clam shell' diagram differs from all of the previous ones. Here, any temporal reference has been removed, and the whole symmetric set of circles implied by the film clip is present at once. The shading of the inner circles suggests a perspective view of the clam shell, with the shaded parts being further away (and hence smaller?). As with the other uses of perspective, this one functions as a boundary-drawing apparatus because of the way that the drawer locates a space between herself and the drawing. It is worth noting that this student introduced written language into the diagramming process as a way of naming and categorizing distinct perspectives. Doing so reclaims the diagrams as forms of representation, and thereby subjects them to the linguistic domain of naming. This multi-modal move is made for clarity in communicating the meaning of the parts of the diagram. And yet there is a sense that the gestural diagramming in this example exceeds the textual naming alongside it, a sense that the embodied hand is still present and no longer silenced by the sayable and the linguistic. It is as though the diagramming is not entirely tamed by the tags, but rather erupts from the page and leaves the text behind.

Inventive diagramming in the classroom

We have focused here only on the diagrams that five students created in response to the prompt. Although we did not videotape the lesson, we did observe several students in the class using their arms in preparing to create the diagrams as well as during the process of drawing. They start with arms held above their heads, fingertips touching, and then separate their hands and circle their arms out until they reach a horizontal, straight position before curving them back towards each other, finally touching them together at stomach height. Incidentally, this set of gestures most closely resembles the last two examples, in which the two fixed points are absent and the line of tangency remains invariant. We have chosen here to focus on the diagrams – instead of also analyzing those gestures – as the locus of the gesture/diagram entwinement. While the arm motions just described offer near-exact representations of the film clip, we were more interested in the way that the students use these visible and kinetic experiences to express time and motion on the two-dimensional pieces of paper. The fifth example, in particular, hints at the ways in which the diagrams might give rise to new gestures that differ significantly from those first evoked by the film.

Another reason for focusing so specifically on the diagrams is to support our investigation into the ways in which the students' diagramming might go beyond – and not just represent – what they actually saw in the film. Thus we opted for the formulation of the prompt 'How does the circle move from being concave up to concave down?', instead of, say, 'Draw what you saw in the film.' In this exploration, we found five different techniques for communicating the temporal, mobile dimension of the film: successive framing, dotted lines, perspective, arrows and shading. The successive framing takes the temporal dimension literally and, due to its discrete character, is less successful in communicating the continuous transformation of the circle over time. Even though both the first and second examples employ this technique, the latter example deploys the spatial arrangement – as well as the arrow – in such a way as to express the whole event as a single story, in contrast to the two separate motions implied in the former example. In the latter case, the straight line situation is seen more as a passing, continuous case, than as a rupture from one concavity to the other.

In the third example, the dotted line is deployed as a way of overlapping the temporality into a single snapshot. The dotted line arcs appear as virtual passages bookended by the actual, static circles that begin and end the transformation. There is a certain continuity expressed here, even though it is not the continuity of the film clip. That continuity is correctly evident in the fourth example, which does not privilege any of the arcs over the other – each one is as real as the next. However, it moves towards a perspective-taking, in which the circles are seen as three-dimensional hoops rotating around an invisible horizontal line. Finally, in the fifth example, the arrows appear as new devices for gesturing towards time and space. Additionally, the shading of the clam shell uses perspective drawing techniques to evoke motion as receding into the third dimension.

We can see in some of these diagrams precisely what Châtelet found in significant historical developments in mathematics: inventive 'cutting-out' gestures that interfere and trouble assumed spatial principles, new and radical 'symmetrizing devices' and the emergence of new perspectival dissymmetries within the given work surface. The fourth and fifth examples are particularly provocative in terms of Rotman's reimagining of embodiment and Châtelet's description of the 'muscular conception of matter'. The move towards the three-dimensional perspective re-images the intangible virtual circles on the screen as material objects (balls or hoops) that can turn – or be turned, with the force of the arrows – around implied spatial hooks and rods. As with Hughes' (1986) diagrams, or those of Bartolini-Bussi and Boni (2003), which include children's hands in their drawings of operating with numbers (reaching for, pulling, holding or touching drawn cubes in the former case, and abacus beads in the latter), these fourth and fifth examples show students moving towards a new mathematical assemblage – carving out a new ontology in the process. Châtelet also offers diagrams like these ones of young children, where the entire body appears on the page with its own orientation that displaces that of the viewer. This introduction of multiple embodied perspectives hails a network 'I' which operates through a plural and distributed agency, forming new assemblages and new folds upon the working surface.

The film clips strike us as especially interesting in that they are essentially virtual and non-tangible – unlike counting beads, blocks or abaci – and therefore not that different from the mental images one might produce in one's imagination. Not unlike Châtelet's description of Einstein choosing to become a (virtual, imagined) photon, so that he could occupy the body-syntonic position of its trajectories, these students include themselves in the spectacle of the moving circle, projecting themselves into the movement, and perhaps even feeling the breaking away of their hands as the circle curls out and stretches into a straight line.

Diagrams and the exteriority of thought

The work of Châtelet challenges educators to reconsider the power of student diagramming as a disruptive and innovative practice – one that sheds light on the very nature of mathematical agency. Such a philosophical shift demands that we examine student diagramming as a gestural intervention into and onto the material surfaces that define our spatial experiences. This is not to dismiss the necessity of developing standard diagrammatic skills for effectively communicating in mathematics, nor to diminish the contribution of research that aims to study how students develop such skills. In fact, the diagramming examples in this chapter contribute to this research by pointing to particular strategies – the use of dotted lines, arrows, rotational gestures, multiple perspectives or points of view, and cut-out gestures that break through or fold the given surface – that are often the mark of enhanced diagramming skills. We have argued, however, that these strategies do not constitute a semiotics that is divorced from the event, but rather a highly material process of becoming entwined and enfolded with the various materialities involved in the assemblage. It is precisely these encounters that we believe substantiate an embodied mathematical assemblage.

In focusing only on the student drawings (and not video recordings of hands, faces, voices, etc.), our aim was to test the interpretive power of these new theories of embodiment in tracking the gestural in the diagram itself. In other words, we wanted to study the extent to which diagrams could be construed as conjuring gestures. This approach allowed us to identify more accurately those particular aspects within the diagrams that pushed at the enveloping gaze of the hinge-horizon. This approach also matched our attempt to treat the diagram as a site of agency and to honour the 'exteriority of thought' while unsettling the inside/outside distinction of Kantian-based theories of the mind (Roth, 2010). We have shown here how the mathematical subject comes into being (is always *becoming*) as an assemblage of material encounters. This assemblage is neither trapped inside an individual body nor confined to a Kantian interiority of unified structural faculties, but rather it is differentiated, heterogeneous and distributed across multiple surfaces. It is in this sense that we embrace the notion of the 'exteriority of thought'.

4 Inventiveness in the mathematics classroom

In this chapter we build on the 'pedagogy of the concept' introduced in Chapter 2 and the diagram-gesture assemblage discussed in chapter 3 to study how learning entails the creation of something new. In keeping with our ontological commitment, the creation of something new does not simply mean the acquisition of a new concept by a learner or a group of learners. Nor does it simply mean participating in a set of practices or discourses, which tends to ignore the mobility and virtuality of the concept. This chapter aims to further elaborate on learning as inventiveness, drawing on Châtelet's study of historical moments of mathematical inventiveness, such as Hamilton's quaternions and Cauchy's residue theorem, all of which involved a creation of the new. As our examples are drawn from the school mathematics classroom, they involve inventiveness at a more local level.

Drawing on our inclusive materialist approach, we aim to study creativity and inventiveness as collective processes rather than properties of individuals. We propose a conceptualization of inventiveness that is not bound to the individual's choice or discernment between alternative possible paths. Creativity is not a property or competence of a learner, as suggested by approaches that seek to measure the flexibility or fluency of a child's thinking. It does not exist independently of its exercise. In other words, it is not that individuals are creative or not creative, but rather that creativity flows across the learning assemblage in a somewhat impersonal way. We thus follow Davis and Rimm (2004) by focusing more on the processes of creation than on the products. Also, our sense of creativity

Portions of this chapter were published in Sinclair, de Freitas and Ferrara (2013). We thank Springer for permitting us to reuse sections of that article.

focuses on novelty, which Plucker and Beghetto (2004) argue is one of the two key elements of creativity, the other being usefulness. These two elements are usually considered qualities of the product, and then projected onto the individual creator (and thus, the learner *is* creative by being fluent, flexible and original). In contrast, we shift attention away from the doer and focus on the doing – thus resisting the temptation to read these actions as reflections of a mental state – in order to enable us to study classroom creativity in new ways.

This chapter explores creativity in the mathematics classroom in relation to both Châtelet's interest in the gesture/diagram interplay and his notion of virtuality, which we discuss in greater depth in Chapter 8. Both classroom examples offered in this chapter centrally involve digital technologies that we think enhance the gesture/diagram interplay in interesting ways. We focus on these technologies both because of their particular relevance to the body-as-assemblage notion we elaborated in Chapter 1 and because of their intimate connection to the notion of virtuality.

Creative acts and materiality

Inventive moments in the history of mathematics, such as those identified by Châtelet, tend to be well-defined, in that they bring forth new concepts that change existing ideas, methods or claims within the larger community of practicing mathematicians. We basically take inventiveness in the mathematics classroom to be similar in its offering of new concepts, but given the relative nature of the new, we need to be more precise about what counts as instances of inventiveness in the mathematics classroom.

Châtelet (1993/2000) emphasizes the materiality of mathematical invention by carefully tracing the life of the mathematician's hand. The making of a mathematical diagram is a material process that precedes formalism and acts as a kind of midwife for implicit, intuitive and even irrational thought, and – in André Weil's (1992) words – for the 'obscure analogies, murky reflections, furtive caresses and inexplicable tiffs' (p. 52) that animate mathematical activity. These diagrams are born out of the mathematician's gesture, which 'cuts out a form of articulation' (p. 8). Diagrams, as we discussed in Chapter 3, are thus conceived as inherently gestural and grounded in the movement of hands. Châtelet's interest is less in the fixed, representational diagrams that eventually get codified in textbooks than in the sketches by which mathematicians create new spaces (new dimensions, new kinds of planes) on a piece of paper, with and through their hands. Such sketches are more like physico-mathematical beings, in that they are not intended to *represent* abstract objects. Mathematical inventiveness, according to this approach, exists in the dance between the gesturing and drawing hand, which expresses and captures the temporal and dynamic moment when the new or the original comes into (*in-venire*) the world at hand.

The inventing of new concepts is often taken to involve a process of abstraction and/or generalisation. But for Châtelet, such an assumption is precisely what leads people to dematerialize mathematical concepts – to see them as mental entities, Platonic truths or as mere discursive arrangements, as we pointed out in our discussion of theories of embodiment in Chapter 1. To avoid this Aristotelian slip, Châtelet relies on the notion of the virtual, which he uses as a way to argue that the mathematical and the physical need not be entirely distinct. Mathematical concepts are thus actualizations of the virtuality of the material world – rather than reductive abstractions from the material world to the immaterial world. Châtelet's notion of the virtual is powerful in large part because he has developed it specifically in the context of mathematics, where questions of the 'concrete' and the 'abstract' are so slippery.[1] As we show in this chapter, the virtual/actual relationship is quite different from the abstract/concrete. The boundary between the virtual and the actual is constantly shifting and being remade in mathematics classrooms through material interactions. In this sense, a creative act can be seen as an ontological reassembling and reconfiguring of the world, a means by which the new comes into being, forever changing the relationship between the virtual and the actual.

From this point of view, we propose to conceptualize mathematical inventiveness in terms of four characteristics, which were inspired by Grosz (2010) and her insightful study of freedom.[2] These characteristics are not

[1] Wilensky (1991) puts forward a strong argument that, contrary to the traditional, largely Piagetian view of what it means to be concrete and/or abstract, concreteness is highly subjective, depending on 'the quality of our relationship with the object'. He discusses the way in which digital technologies can enable qualities of relationships that make hitherto abstract (and much-hated) concepts such as fractions feel very concrete.

[2] Grosz draws primarily on the ideas of Henri Bergson to rethink freedom and agency. Pimm and Sinclair (2009) also draw on the notion of a 'capacity to do things', based on Ober's (2007) assertion that it is the defining trait of democracy, to explore the question of who has, or who can assert, the power to do things in mathematics and on what basis.

prescriptive, but they honour the indeterminacy that fuels ontogenetic processes of becoming inventive.

A creative act:

1. *introduces or catalyses the new* – quite literally, it brings forth or makes visible what was not present before;
2. *is unusual*, in the sense that it must not align with current habits and norms of behaviour;
3. *is unexpected or unscripted* – in other words, without prior determination or direct cause;
4. *is without given content*, in that its meaning cannot be exhausted by existent meanings.

The first characteristic pertains to Châtelet's process of actualizing the virtual. This is an ontological claim about what constitutes the new. In actualizing the virtual, a creative act brings forth – literally makes manifest – an object which did not exist prior to the act. The second characteristic attends more carefully to the specific social context where the act occurs, and thereby frames the act as creative in relation to particular practices that are taken as norms. Thus, creative acts are deemed as such in relation to governing norms; the extent to which they are recognized as creative is conditioned by the context in which they occur. The third characteristic points to the collective emergent nature of creative acts, whereby the new arises without being directly and formally determined by the intentions of the individuals involved. And finally, the fourth characteristic underscores the ways in which creative acts change the way language and other signs are used and alter the meanings that circulate in a situation. Indeed, creative acts bring forth new uses of language and often break with the rules of common sign use, so that the new can be distinguished from that which is already familiar. These four characteristics point to the centrality of the body and its movement in acts of creativity, rather than focusing on internal mental dispositions and intentions.

In the classroom situations we discuss in this chapter, mathematical inventiveness is considered to be a relationship between learners/teachers and the material world. This allows us to resist the tendency to locate learning in an individual body and, instead, to consider the ways in which learning is distributed over a collective assemblage. Here we follow our body-as-assemblage approach from Chapter 1, in which the body is reconceived in terms of distributed agency across a network of intra-actions, the properties of which are constantly changing. The particular context of the technology-based classroom features digital tools that participate in Rotman's network 'I' by decentring human agency and radically transforming the possible arrangements at work in the classroom. The potential multiplicities implicit in any of these assemblages can be awakened by material actions (gestures and diagrams) that constitute inventive moments (processes of actualization). Before examining that which affords such an environment its creative impulse, we elaborate on the decentring and transformative features that we see digital technologies bringing to the mathematics classroom.

Digital technologies, mathematics and impulse

Although the word 'virtual' is often associated with the computer, we seek to remain faithful to Châtelet's use of the term, which has no digital requirements. That said, we think there are features of certain digital technologies that make it particularly conducive to creative acts. Since its exciting and eye-opening beginnings with Papert's (1980) Turtle Geometry, scholars have touted new digital technologies as being capable of radically changing the way students think and learn. Of particular interest to us are the so-called *expressive* technologies that provide tools that enable learners to construct mathematical objects and explore relationships among them. The body syntonicity of Logo carves out a new subjectivity for the learner, who is now at the centre of the mathematics and *is* the mathematics. A mathematical square becomes a means of moving within a vocabulary of walks and turns that is in stark contrast to the square as a particular visual configuration or a particular property-based definition. We see such square-making experiences as potential inventive moments in which the human-technology assemblage gives rise to new ways of thinking and moving. Similarly, data collection and physical output devices (e.g., MBLs or motion detectors) have introduced into education significant ways to connect simulations and real phenomena. They eliminate the algebraic as the sole channel into mathematical modelling and enable for learners a challenging active *kinaesthetic* engagement with the technology (see Ferrara, Pratt, & Robutti, 2006).

In this chapter, we focus on two particular digital technologies that both attempt to mobilize mathematics: dynamic geometry environments (DGEs) and motion detectors. Unlike Logo's more static drawn surface, these digital technologies temporalize mathematical behaviour. In a DGE, for example, a triangle is not a representation of the abstract triangle, nor is it an example of a particular triangle; rather, it is *all* and *any* possible triangles, which the user actualizes by dragging the vertices that determine it. The triangle has been inscribed in a new space, a stretchy space of continuous transformation. In describing the shift from declarative geometry (written proofs and command-driven constructions) to dragging geometry on one's screen with the mouse, Jackiw (2006) writes, 'one's actions are inquisitive and usually tentative: one is seeking, rather than stating' (p. 156). The seeking hand, and – with some motion-controller technologies, the seeking body – can not only move freely across the screen, but they can also put into motion an 'improvisational choreography' of mathematical objects, with the trajectory of one object mathematically dependent on others. The result is 'a single possible performance drawn from the limitless configuration space of the mathematics spread across the stage' (p. 156). Jackiw's words are profoundly important in understanding the possibilities of the gestural/diagrammatic interplay, precisely because of the way in which the seeking hand – which is tentative and awkward at first – learns to move.

In the case of motion detection devices, the real time feedback of the tool is what makes the graphs on the screen dynamic and responsive to different kinds of motion performed with either the user's body or an object. New ways of thinking are offered through the experience of this sensorimotor feedback (for example, you will move your hand faster when you anticipate steeper graphs; you will imagine, draw and gesture new diagrams as generated by particular motions you have not performed before). In discussing the change provoked by data capture technologies, Ferrara et al. (2006) emphasize the important interplay between the physical actions of the students and the real time appearance of the graph on the screen; not only does the graph capture the mobility of the students, but as students see what happens on the screen, they can change the way they move. There is a double sense of controlling and being controlled in this simulation. Nemirovsky and Ferrara (2009) illustrate such gestural/diagrammatic interplay in their description of one girl's gestures tracing the motion of two laser lights in order to discover a defined triangle shape that gives the trajectory of the composed motion.

The virtual is actualized in large part by the fact that, in these environments driven by the hand or body, humans are constantly reinscribing themselves into the idealized, abstract mathematics. Speaking specifically of dynamic geometry (although his ideas are also applicable to other dynamic technologies), Jackiw (2006) writes that it is a milieu in which 'the individual "touches" raw mathematical ideas, where personal volition and physical exertion can make seismic impact on disembodied abstractions' (p. 155). Of particular interest here is the materiality of the mathematical objects being 'touched' and the creating of something that is more than the sum of its parts. In the next sections we present two case studies involving young children working in mathematical environments that feature the use of digital technologies. Our goal will be to illustrate the characteristic features of inventiveness proposed earlier in this section, which we do by highlighting the material entanglements in these mathematics classrooms.

When do two lines intersect?

The episode described in this section involves grade one children seated on a carpet in front of a large screen. The classroom teacher sits on the carpet with the children, and the instructor (Sinclair) sits at a child-sized table with her computer, which is running a DGE called *The Geometer's Sketchpad*, so that she is facing both the children and the digital projector. The investigation focuses on lines and, especially, on how two lines relate to each other. The children had not formally encountered geometric lines, or the notion of intersecting lines, previously.

The lesson begins with the children being shown several examples of pairs of points tracing out thickly coloured linear paths, with some pairs intersecting and others not (see Figure 4.1). In talking about these pairs of lines, the children describe the former as 'touching'. After the children successfully identify pairs of lines that 'touch' and those that do not, the instructor offers the more technical word 'intersection' to describe the former, which the children immediately connect to road crossings and car crashes.

The instructor opens a new sketch and uses the line tool to construct two lines, colouring one red and the other blue. The lines are positioned

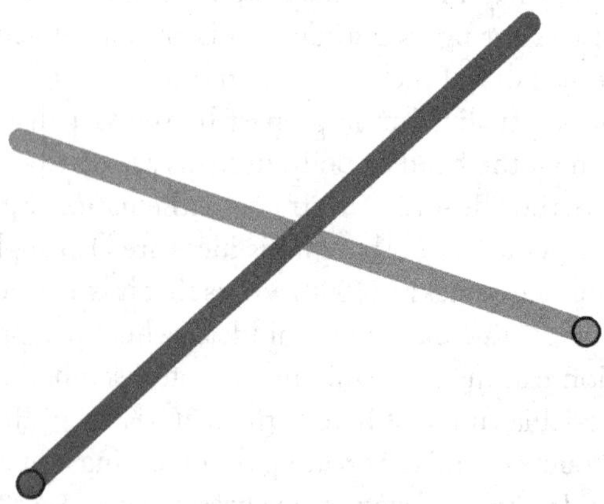

Figure 4.1. Two points tracing intersecting paths in Sketchpad.

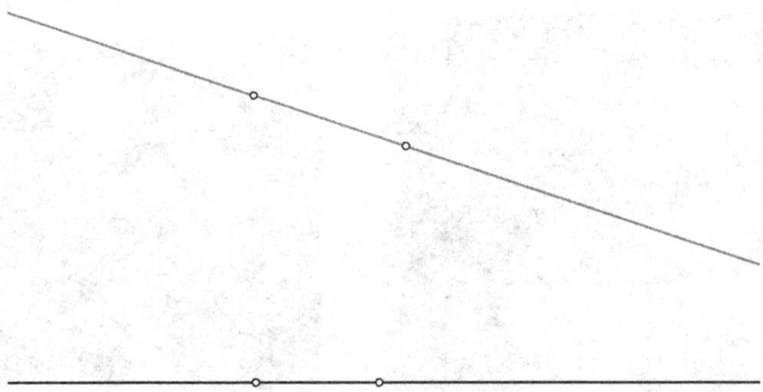

Figure 4.2. A non-visible intersection in Sketchpad.

so as to be non-parallel, but so that the intersection is not visible (see Figure 4.2). When asked, 'Do you think these two lines meet?', the children all say 'No' in chorus. Then one girl says, 'But they can if you tilt it all the way down.' The instructor begins dragging the top line towards the bottom one and, as the intersection becomes visible, one student says, 'Now they have an intersection' – a temporal observation suggesting they did not have one earlier. Another adds, 'a very small one'. The instructor drags the top line up again to its original position and asks, 'And here do they make an intersection?' The children chorus, 'No.' After a few seconds, one boy says, 'Oh yes they do, they do.' Several children begin talking at once, and one says, 'Because they go out of the screen.' So the instructor adjusts the screen (dragging the right corner of the window to enlarge it), and an intersection becomes visible.

The instructor then drags the lines even further apart, so that their intersection is not visible, and she asks the children, 'use your imaginations' to decide whether they intersect. This time most children say, 'Yes.' Then a few say that they would not, with one girl explaining, 'because they are very far apart'. Other children hedge they "think it might'.

T: Can we make some theories about why it might intersect?

Natasha: Because it's tilting (*referring to the red [top] line*).

The instructor invited other children to explain their reasoning.

Robert: The lines, um, can't meet at the edge of the screen because they are too far apart (*left hand raised with index finger and thumb forming a 'C' shape*) and they can't just like suddenly just have a straight line going down and meet (*index finger and thumb coming together, Figure 4.3a*).

Jamie: Cause they are going like this (*two arms moving along a linear path, Figure 4.3b*).

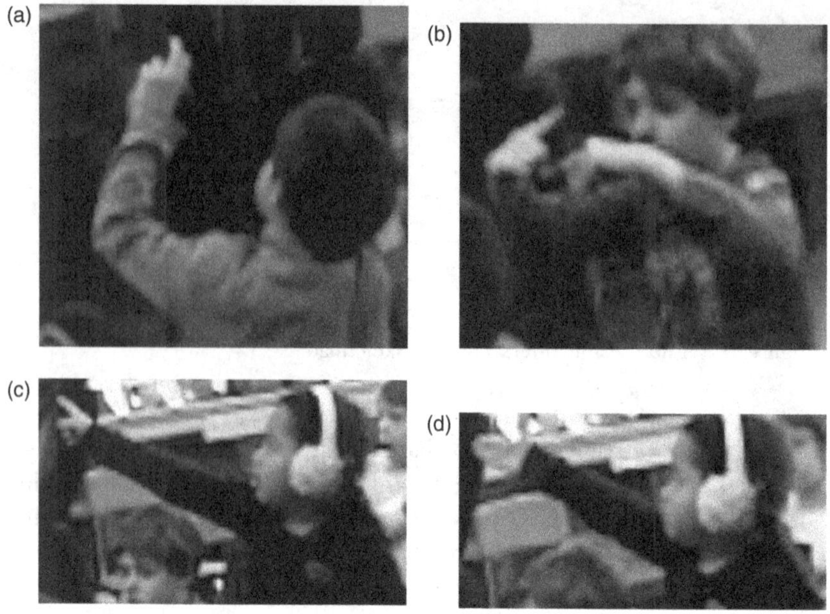

Figure 4.3. Children's gesturing with the lines.

T: But do you think they would ever meet?

Robert: Yes, because they are both slanting and the red one is slanting toward the blue one.

The instructor repeats Robert's reasoning and then invites more contributions.

Natasha: It's going to always connect somewhere because the red one is slanting (*tracing index finger along a linear path, Figure 4.3c*) so it's going to connect somewhere over here (*having moved hand to end of screen, turning it into a vertical position and moving it up and down, Figure 4.3d*).

T: Even if we can't see it, it's going to connect, it's going to intersect somewhere over here?

*Jamie:*I think it's never going to intersect.

T: Why?

Jamie: Because I just do.

T: What do you think about the theory though that this (*pointing to the red line*) is slanting more and more toward the blue?

Jamie: (*Standing up*) But the blue is also going like this (*using hands and arms to show that both lines are slanting, Figure 4.4a*).

T: Oh I see. Interesting, so the blue is slanting as well.

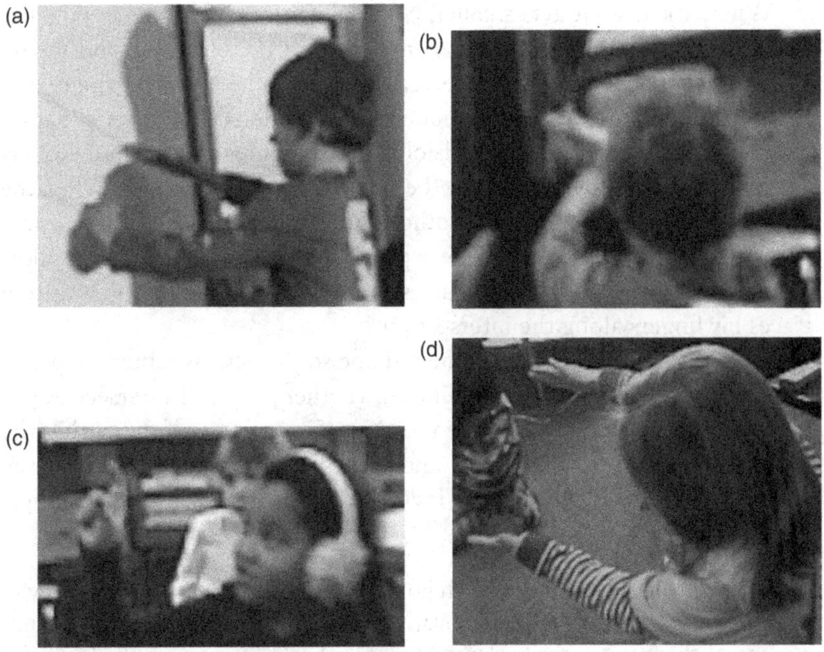

Figure 4.4. Children's gestures evoking new objects.

Jamie: As long as both, the red's going down the blue's going down beside it so the line can't just go like that (*bringing his hands together, the top one curving down to touch the bottom one, Figure 4.4b*) and then intersect.

T: That's interesting. Let's look at a situation where we can definitely see an intersection (*dragging the two lines so that their intersection is visible on the screen*). So now they're both slanting just like Jamie said before.

Natasha: But it's always going to slant because right there (*pointing to the left line on the screen*) that's how thick it was so it's always going to slant.

T: It's always going to slant.

Saskia: It's going to intersect.

Robert: It's going to intersect at one point but it might, it might intersect somewhere far, far away.

T: We need to figure out how we're going to know when the lines are going to intersect even when we can't see it. So Jamie, no Natasha, said they're going to intersect because the red one is slanting toward the blue one.

Natasha: No, because that right there (*hand positioned so that index and thumb at a certain distance away, Figure 4.4c*) isn't the same thickness and it's going to always intersect because it always gets smaller.

When asked what gets smaller, Natasha comes to the screen, puts her index finger on the red line and her thumb on the blue line, and moves towards the intersection while decreasing the gap between her finger and thumb. The instructor then announces they will look at another situation in which the intersection is not visible. After dragging the red line, Robert asserts that the lines will intersect 'because it's slanting enough'. When the instructor proposes to look at another one, Jamie asks, 'Can we see if it is going to intersect or not?' No one expresses any surprise when the window is enlarged in order to make the intersection appear. Jamie then gets up and traces his fingers along the intersection.

Finally, the instructor drags the red line so that the two lines are parallel to each other and asks the children whether they will intersect. All of the children say, 'Nooo.' Camille uses Natasha's gesture of measuring the thickness. Jamie uses both arms and says, 'because they are going away from each other'. The instructor invites a student who has not spoken yet to contribute:

> *Charlotte:* Because they are both going the same way. One of them, they're not slanted, so, they're kind of slanted but they're not going to meet since one of them is not really slanted because they're just going like (*gesturing with one straight arm the direction of a line*) they're both going (*now bringing the other arm to move parallel with the first*) like that so they're never going to meet (*right hand curving down towards the left one, Figure 4.4d*).

The instructor then offers the word 'parallel' to describe two lines that are never going to intersect.

Creating a new space for potential intersection

Two strategies are collectively generated for solving the problem of deciding when two lines intersect: (1) the idea of the lines intersecting because one is slanted more than the other (or is slanted enough); and (2) the idea that the lines intersect because the thickness between them is changing. Gestures are used throughout as the children make arguments about what will happen to the lines. The first gesture by Robert shows that the lines are 'far apart' and they cannot suddenly 'meet' at the edge of the screen. Interestingly, Natasha's gesturing of 'thickness' also relates to the distance between the lines, although hers is one that she describes as being able to change over time. However, before Natasha talks about thickness, Jamie and Natasha use their hands and fingers to invoke the current and future paths of the lines. Jamie's hands *are* the lines, moving steadily from left to right, whereas Natasha seems to point to the path of the line on the screen. Jamie's use of arms-as-lines is later used by Charlotte to explain why the lines will never intersect. We see these gestures as being evoked by the dynamic tracing out of lines they saw previously: Lines are all traced out over time and are not just represented by, say, static arms placed at an angle to each other. This evocation of the lines is precisely what enables the movement past the limits of the screen and enables the children to create the possibility of an intersection that is not visible beyond the objects on the screen. If an intersection was something concrete and visible earlier, it later becomes something that can be imagined, potentially existing by virtue of the gestures that sustain the relative slanting or changing distance.

If the gesturing of extending lines brings to life the invisible intersection, the 'thickness' gesture invokes a new relationship between the two lines, that of distance. First used by Robert to explain why the lines could not meet at the end of the screen, Natasha uses it again after seeing the screen scroll in order to make the intersection between the lines visible. In scrolling the screen, the lines themselves remain static, but the 'thickness' changes. It is this changing quantity that Natasha recognizes. Again, this gesture is later used by Camille to describe the invariance of the distance between two lines that will never intersect. What is also of crucial significance in this gesturing of thickness is how it entails attending simultaneously to the moving 'image' of the lines *and* the material surface on which they are projected. This dual and paradoxical nature of visual perception – that we simultaneously see both an image and its medium – operates beneath perception at the level of sensation. As we discuss in Chapters 6 and 7, sensation and micro-perceptions are like the glue that keeps particular learning assemblages together. In carving out the thickness between the lines, new material entanglements are forming between the emergent concepts and the white empty space between the lines, a space that is attached to the screen in a different way than the lines themselves. In other words, conjuring this thickness indicates how the children are animating both the lines and the empty space of potentiality where the lines live.

In summary, we see this episode as involving a series of gestural and verbal thought experiments that eventually unleash the potential point of intersection. The creative act involves the slow expansion of the plane circumscribed by the screen, extending it beyond what was previously visible to a plane that can welcome the crossing of lines not drawn. The potential point emerges in both the interactions amongst the children and their interactions with the screen/software (which sometimes shows the intersection and sometimes does not, but always maintains the line as perfectly straight yet infinitely variable in the slant of that straightness). This reading of the episode focuses less on the creativity of any given child and more on the unexpected interactions between the material and human players in the classroom. In terms of our four-fold characterization of creativity, we claim the following:

1. There are several creative acts in this example. The first is the extension of the surface of the plane, which literally brings forth or makes visible that which was not present earlier and unleashes the potential point of intersection. The technology plays a central role in affording this material act of creation. In addition, the children perform creative acts in gestures that literally make manifest the convergence and intersection of the lines. This catalyses two ways of explaining when two lines will intersect in cases where the point of intersection is not visible.
2. Given that the norms of behaviour in the classroom in relation to lines and planes involves working with the concrete and visible, the collective actions (both movement and discourse) by which the plane is extended and the point of intersection is created can be considered unusual, because such actions involve the non-visible and the potential. One could also argue that the particular gestures deployed by the children are unusual and break with gesture norms, although we are unable to say so definitively without more data. It is evident, however, from facial and other expressions, as well as from the teacher's invitation to repeat the gesture, that Jamie is creating and using gestures in ways that are entirely new to him.
3. The creative acts are genuinely unexpected and unscripted in the sense that the teacher is experimenting with a new technology, as well as with ideas that are not usually part of the grade one curriculum. More importantly, the creative acts are also unexpected for the children. This is important because the teacher needs to be able to occasion similar creative acts with other groups of children. But even more importantly, the creative acts are unexpected in the sense that they are not *directly caused* by the software, by the teacher or by any individual student.
4. The existent meanings for 'line' and 'intersection' defined them in terms of their concrete and visible nature. The unfolding path of the lines on the screen, as well as the 'uncovering' of a hidden intersection, provoke gestures amongst the children that actualize infinitely extending lines and their invisible points of intersection. The

> new meanings of 'line' and 'intersection' are by no means exhausted by the old ones, in the sense that the shift from the possible objects on the screen to the potential ones travelling off and on the screen fundamentally changes their nature.

We have purposely refrained from ascribing creativity to any one individual. Instead, in addition to the chorus of words and gestures that circulate in the classroom, we highlight the projected dynamic diagrams, as well as the computer and the instructor, in the collective and creative activity that furnish the virtual space for the invisible intersection point and catalyse new gestures and meanings for the children.

What kind of motion makes a vertical line?

The episode described in this section was part of a larger research project aimed at studying the potential for a graphical approach to functions through the aid of motion detectors in grades two through five (see Ferrara & Savioli, 2009). The particular lesson described here was conducted in a regular grade four classroom and focused on different models of motion. It begins with the teacher asking the children to recollect the previous grade three explorations they had done with the software *Motion Visualizer DV* (MV). The software works through the aid of a web camera linked to the computer. Based on live input, the MV captures and tracks, in real time, the motion of a coloured object in a plane (in this case, the object is an orange glove that enables the tracking of hand movements). As the student moves the object in front of the web camera, the software displays – on the right side – two graphs decomposing the motion into the two dimensions of the plane and – on the left side – the trajectory in the three-dimensional space of the room and the live video of the student moving the object on the plane (Figure 4.5).

The teacher (and then the children) refer to the paper on which the movements were performed as 'Movilandia', and the screen showing the graphs, which are generated by the movements of objects across the surface of the paper, as 'Cartesiolandia'. Each graph shows the movement of the object in relation to the particular dimension (in this case, horizontal and vertical dimensions). In grade three, the children had moved the glove in Movilandia along straight trajectories – horizontal, vertical and oblique – and had watched the corresponding motion graphs that were generated on the screen, as well as the graphs that were generated when the glove was kept still, in each case investigating the associated relationships between position and time.

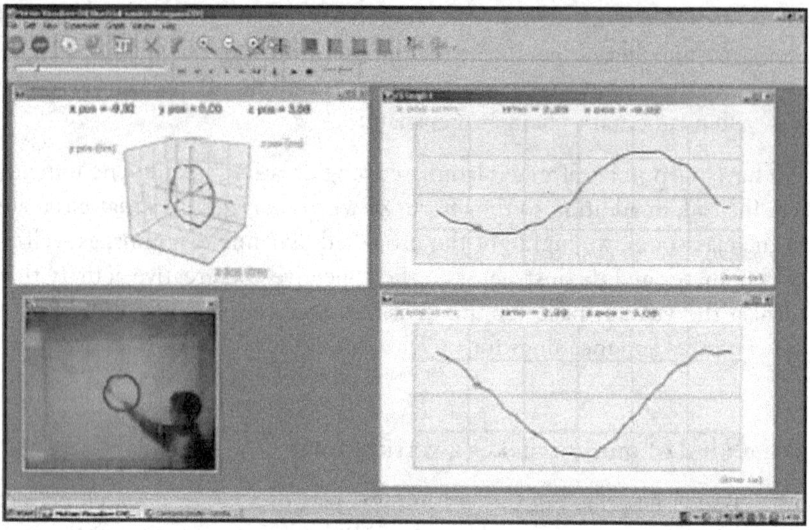

Figure 4.5. Interface of the Motion Visualizer.

The following episode occurred in a lesson aimed at reflecting upon the various kinds of motions already experienced in experiments with the MV. It begins with the children being asked what they remember about past activities with the MV, starting with 'shapes' in Movilandia, that is, motion trajectories. The discussion occurs in the classroom, but without the presence of graphs or screens. In talking about how many 'shapes' had been seen in Movilandia, several children recall three main, straight trajectories; one child summarizes them as 'oblique, vertical and horizontal'. At this point, the teacher perceives a latent confusion between the two worlds of Movilandia and Cartesiolandia, and invites the children to discuss the case of the oblique trajectory:

> *T:* Oblique, how?
>
> *Arianna:* It is made in a diagonal (*left hand raised and moving in an oblique line*) that, when you move in Movilandia, in Cartesiolandia, um, you move in a certain way in Movilandia and a line, vertical or horizontal or oblique, appears in Cartesiolandia.
>
> *Elisabetta:* But it depends on the way you move.

At this point, Marco immediately points out the impossibility of having a vertical line appear in Cartesiolandia: 'It cannot be vertical (*speaking in a whisper*).' The teacher repeats Arianna's words, including her reference to a vertical line: 'You said that a line, vertical or horizontal or oblique, appears in Cartesiolandia.' Marco repeats his objection and then makes a distinction between what can happen in Movilandia and what can happen in Cartesiolandia.

Marco: No, a vertical line never appeared in Cartesiolandia.

T: Did a vertical line never appear?

Marco: In Movilandia we moved along vertical segments, but in Movilandia (*sic: meaning Cartesiolandia*) a vertical line never, it never appeared.

Elisa: Or in a diagonal direction, um, or in a horizontal direction.

The teacher invites the children to explain the fact that a vertical line never appears. In the following discussion, three children – Gaia, Elisa and Beniamino – give different explanations, shifting back and forth from talking about what had happened to what could happen.

T: Why could not a vertical line appear?

Gaia: Because the glove, um, when it moves, it moves (*right hand raised in the air miming a short movement*) from bottom (*right hand closed in a fist indicating a specific position*), since, um, when it appears in Cartesiolandia, the glove is always at the bottom (*indicating a specific position*) and then it makes the line in this way (*right hand shifting horizontally from left to right, Figures 4.6a and 4.6b*) as they moved, and it does not start in this way (*right hand moving twice along a vertical direction, from top to bottom, Figures 4.6c and 4.6d*) to make, um, the vertical line.

T: What do you want to add, Elisa?

Elisa: To me, because in the table (*left hand kept still in the air, right hand miming the axes*) that is in Cartesiolandia, it appears, um, to come vertical, it does not arrive at the end of the table (*open right hand moving horizontally, from left to right, Figures 4.7a and 4.7b*), but it should arrive at the end (*right hand repeating previous gesture, Figures 4.7c and 4.7d*).

T: What about you, Beniamino?

Beniamino: I wanted to say that, as Elisa said, there is the table (*left arm raised vertically*), where here there is time (*right hand moving twice horizontally direction, from left to right*) and here (*left hand shifting twice vertically direction, from top to bottom*) there is the movement you make, um, but you cannot, for example, in little time, say, 10 seconds, in few seconds make, um, be able to have such a movement (*left hand miming a vertical line*) on a platform, that is in a place making you understand that time passes (*left arm raised vertically, and right hand moving horizontally from left to right, Figures 4.8a and 4.8b*),

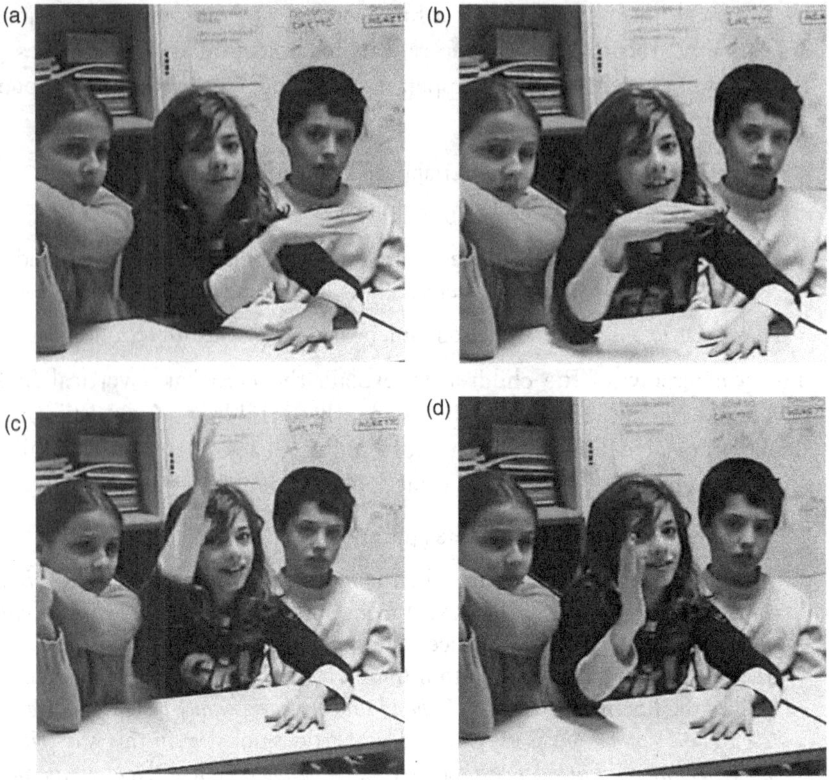

Figure 4.6. Gaia's right hand moving horizontally and actualizing the vertical line.

since it would be as if you stopped time (*left hand pointing to a specific position, Figure 4.8c*) and moved (*left hand jumping twice in the air, Figure 4.8d*).

The teacher then helps the children share this 'as if' movement in the classroom and translate it in a straightforward relationship between variables in the graph.

T: If you stop time, it is as if time didn't change, but what does change?

Beniamino: Um, the movement.

T: The movement?

David: The position.

T: In what you call the table, what appears vertically?

Ss: The position!

(a) (b) (c) (d)

Figure 4.7. Elisa's right hand moving twice horizontally.

T: So, to have a vertical line (*right hand miming it*) it should be, um, I stop time but?

David: The position changes.

Ss: Yeah (*laughing*).

Elisa: So, it's impossible (*with emphasis*)!

In this episode, the new idea of the vertical line as a model of motion emerges. Arguments about what would happen in this instance are driven by hand and arm gestures. The discussion is initiated by some statements by Arianna and Elisabetta that seem to indicate confusion on their behalf about what can occur in Cartesiolandia. Marco immediately states, as a correction, 'It cannot be vertical, and Elisa and Gaia attempt to explain why it cannot be vertical. Together they point to how

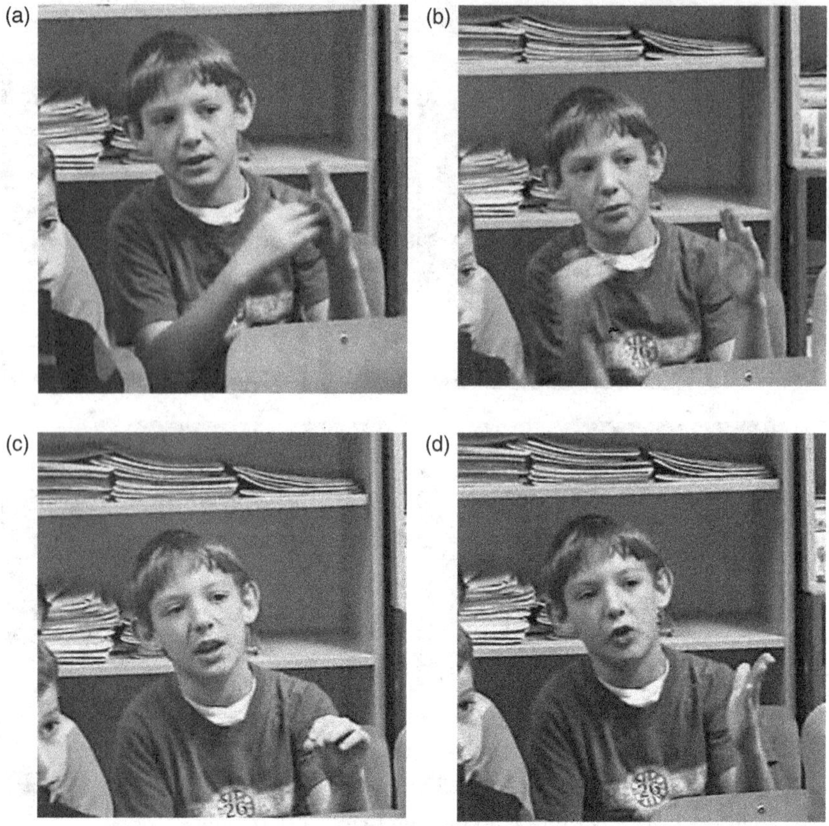

Figure 4.8. Beniamino's right hand miming the passage of time and his left hand actualizing a timeless motion.

time starts at zero (Gaia) and ends at some positive value (Elisa), and thus all graphs in Cartesiolandia, regardless of their shapes, begin at the left and end at the right along the horizontal axis ('when it appears in Cartesiolandia ... it makes the line in this way as they moved'). For Gaia and Elisa, it is the flowing or duration of time – and the fact that time starts and stops – that is used as an explanation. Although the vertical line had never appeared as a graph of Cartesiolandia earlier, it can now be invoked by the right hands of Gaia and Elisa, as they move their hands vertically up and down. As their hands move, however, they also specify the impossibility of such a graph in Cartesiolandia: 'and it does not start in this way to make, um, the vertical line.' This reveals the *power* of gesture to conjure (and in our terms, create) an entity that is self-contradictory.

Elisa picks up on Gaia's newly conjured entity, exploring the consequence of 'to come vertical', and then what would happen 'at the end of the table'. By referring to the table, she draws attention to the axes of the graph and their relation to time and position, which Beniamino recognizes. When he says, 'you cannot, um, be able to have such a movement on a platform, that is, in a place making you understand that time passes', Beniamino is asserting that the motion that would generate a vertical line in Cartesiolandia cannot be a *real* motion in Movilandia. This assertion is based on the claim that the glove would be at different positions at the same time. However, once again, the fact that the motion cannot be realized in Movilandia does not prevent its actualization through the gestures, with Beniamino's left hand pointing to a specific position in the air ('you stopped time') and jumping from left to right ('and moved'), specifying that 'it would be as if you stopped time and moved.' His argument is thus quite different from the ones offered by the two girls. Rather than referencing the invariant aspects of graphs in Cartesiolandia, like Gaia and Elisa do, and appealing to the experiential fact that time's arrow moves in a positive direction, Beniamino points to the *impossibility* of moving without time passing. This argument introduces a counter-factual into the mix. The experience is so immersive that Beniamino uses the 'as if' form and the subject 'you' (a generic 'you').

We see in this episode an interesting case of collective creativity where the distinction between the possible and the potential is explored through a gestural rendering of an impossible graph. In other words, the children's gestures of a vertical line do not correlate to any legitimate mathematical entity in this context, but they nonetheless allow the children to explore the question as to the logical possibility or impossibility of a vertical line. If earlier the vertical line had not at all been present in Cartesiolandia, it later becomes part of the gestural potentiality of Cartesiolandia, something that can be imagined and manipulated. Moreover, the children begin to tap into the potentiality of this line through gestures that operate outside the domain of the logical. Gaia, Elisa and Beniamino make a series of gestural and verbal thought experiments with the hypothesis of the vertical line as a potential graph. The creative act involves the actualization of a motion that does not move – the reconfiguring of space-time relations so that the glove *might* (but does not) occupy two distinct positions at the same time. The virtual vertical line emerges in the lived contraposition between the real experiments, their possible models and potential unscripted futures, and a recollection of past experiences with the MV and the graphs it produced.

In terms of our four-fold characterization of creativity, we can say:

1. The actualization of timeless motion through the children's hand gesture, which literally brings forth or makes visible an object that is outside the realm of the possible, is a creative act. This shifts the boundary between the virtual and the actual, and the related but distinct boundary between the possible and the real. These gestures highlight the difference between creative acts of actualizing and logical inferences that realize the possible. They also point to the role of the absurd in inventive activity.
2. Given that the norms of behaviour involve working with the logically possible, the act is unusual, because it involves deploying gestures that engender a previously non-existent entity – one that is not possible within the MV environment. Moreover, one might simply point to how this collective activity reconfigures the domain of what is mathematically possible and thus breaks with norms of usual behaviour. We want to stress how this is also an episode where gestures redefine what might be considered possible. In this sense, the children's gestures show how the potential animates or destabilizes the possible, which opens up discussions about divergent and creative activity that operates alongside or within the rules of inference.
3. The creative act is genuinely unexpected and unscripted in the sense that the teacher was experimenting with a new technology, as well as with ideas that are not usually part of the grade four curriculum. More importantly, the creative act is also unexpected for the children. This is important because the teacher would want similar creative acts to emerge with other groups of children. But even more importantly, the creative act is unexpected in the sense that it was not directly caused by the software, by the teacher or by any individual student. The creation of a vertical line through gesturing emerges collectively through interaction between the children and the researcher as they recollect previous encounters with the MV. The potentiality of the vertical line emerges from the discussion itself and from the need to understand the differences between student contributions.
4. The existent meanings for 'line' become linked to conceptions of time and temporal flow through the experiments with the MV. Prior to the lessons, the children had not worked with lines as

> material unfolding paths. The discussion shared here indicates that the children's memories of the events were strongly bound up with how the line was a material unfolding in time and space. The unfolding path of the vertical line as a graph generated by some sort of movement provokes attempts at an explanation of its impossibility. The attempts are driven by gestures that actualize consequences of the invisible vertical line and its meaning as a model of motion. In these attempts, we see how the ontological and the logical are completely entangled. This changes the nature of the meaning associated with the graph of a vertical line in an unexpected way and favours a shift from the possible graphs on the screen to a potential new graph corresponding to an imagined movement.

The children create a new space where they can reason about the graph of the vertical line: a gestural space not physically possible, but mathematically actualizable. While Gaia and Elisa keep thinking about the necessary truth that time unfolds or endures, and that there are thus conventions or invariants of graphs in Cartesiolandia, Beniamino's thought experiment shifts attention to the virtual through a movement that happens in 'no' time. This creative act is the seed for the idea – shared in the classroom – that instantaneous motion *cannot* happen, but *it could* happen. The discussion shows how conditional language maps onto the virtual space of potentiality.

Mobility and potentiality

We now wish to consider what kinds of conditions give rise to these two episodes of inventiveness and, in particular, to articulate the roles played by the digital technologies. With respect to the latter issue, we do not believe that creative acts in the mathematics classroom require the use of digital technologies, nor that the use of DGEs and MBLs are sufficient enough to result in creative acts. We are also reluctant to describe the particular technologies in terms of the affordances they offer, if only because we aim to study the distributed agency that circulates across the entire learning assemblage. Keeping this in mind, it is possible to investigate the features of the use of these technologies that enable actualizations of the virtual. Might mobilizations of mathematics – as exemplified here in interactions with technologies that animate diagrams and evoke a vibrant dynamic potential (or virtuality) – open up all sorts of opportunities for creative acts? Burbules (2006) pursues a similar exercise in his attempt to identify the features of digital technologies that may produce the sense of immersion associated with his construct of virtuality. Despite differences with our approach, we find useful his five features of digital technologies associated with producing virtual experiences: mobility, inhabitation, action at a distance, haptic sensitivity and performative identities. All of these features essentially involve the potential: Mobility refers to being able to really move things (lines, points, ourselves) in new spaces (not ones that satisfy our normal physical laws); inhabitation refers to the extension or transformation of space and time, and the bodily occupation of that space and time; action at a distance refers to our ability to transform the temporal dimension of our participation; haptic sensitivity refers to the way in which our bodies are firmly implicated in the virtual spaces we explore, enabling a rapprochement of body and machine that reconfigures sight, touch and sound; and, finally, performative identities refers to the extension and transformation of our identities in cyberspaces.

Both technologies in these two case studies are first and foremost about mobility. But in the context of mathematics, this mobility is even more poignant than Burbules lets on, in part because of the ongoing program of detemporalization that *is* formal mathematics and in part because of the status of mathematical objects being more or less inaccessible to being moved. For the intersecting lines example, the movement of the points and lines occurs in a frictionless, infinitely extendable, two-dimensional space. The children soon join this new world, using their bodies, arms and hands to conjure more lines, thereby extending and transforming their own spaces beyond that of the visible (e.g., 'it's going to connect somewhere over here'; 'it's always going to slant because right there'; 'it might intersect somewhere far, far away'). And while they do not interact directly with the mouse, or even the points and lines (the teacher does the 'dragging'), their bodily involvement is acute, as can be seen in the 'dynasties of gestures', to use Châtelet's phrase, they produce. It is initially important for the children that there be the possibility of moving the screen in order to make visible the point of intersection – here the children use the language of what 'might' happen. And perhaps the shift to the potential was aided by the fact that they did not have direct access to the mouse, which occasioned the bringing forth of shared gestures.

In the second example, we see movement both when a child walks in certain ways so as to create graphs and in the avatar on the screen, which responds to the child's movements. The movement is highly coupled with the sense of action at a distance, as the child brings into being shapes on the screen through the behaviour of his or her body. As with the previous example, the bodily involvement (now in a space where the technological device is no longer physically present) is palpable, as the children use their arms and hands to conjure lines they have seen, as well as lines they can imagine (e.g., 'the glove is always at the bottom'; 'it does not start in this way'; 'it does not arrive at the end of the table, but it should arrive at the end'; 'you cannot ... be able to have such a movement'; 'it would be as if you stopped time and moved'). While their previous work with the technology had begun in the real, the interplay of their mobility and the inscriptions on the screen first leads to possibilities ('I could move this way', 'I could produce that graph') and eventually to the potential of timeless motion (expressed in terms of 'as if').

Burbules' characteristic feature of performative identities, which emerges from his consideration of technologies such as social networks and virtual realities, seems at first to be much less relevant in our examples. However, we follow Rotman (2008) in asserting the way in which mathematical activity co-involves the discipline (and its concepts), the person and the material world – and that this co-involvement means that mathematical activity does not just produce more mathematics (or more learning), but also produces a new person in a new material world. And if we pursue Barad's post-humanist challenge, the notion of performative identities needs not be reserved for the people involved (in this case, the teacher and the children), but applies also to the concepts themselves.

We see these two case studies as illustrating the notion of a 'pedagogy of the concept' that we introduced in Chapter 2. The tasks designed in these examples feature particular kinds of catalyzing concepts – intersecting lines and timeless motion – in a context in which their mobility can be sustained. These concepts resist the determination of knowledge precisely because they function as material and ontogenetic devices in which the new can come into being. In both examples, learning – the creation of the new – comes about from situations that were previously unimagined, impossible, unusual and unexpected: The creative acts collectively engender a new space, which enables new forms of arguments to emerge. As we showed, the diagram/gesture interplay provides a gateway to virtuality. The children's gestures are not windows into deduced or induced inferences; rather, they bring into being new mathematical objects that can be shared, in full sensuous inventiveness, in the classroom.

Drawing on Châtelet, we can see how gesturing and diagramming can together bring about new ways of thinking, moving and imagining, and thereby give rise to inventive processes. By separating the processes of actualization from processes of realization and distinguishing between the potential (virtual) and the possible, a distinction we discuss in more detail in Chapter 8, Châtelet allows us to study the ways that learners and teachers bring forth mathematical entities as material inventions and not simply as logical deductions. In a sense, the concept of the virtual becomes the animating force of the mathematical, giving flesh and mobility to what might otherwise have been considered abstract, idealized and inert.

5 Materialist approaches to mathematics classroom discourse

One strength of Châtelet's philosophy of mathematics is that it draws attention to the crucial role of gestures and diagrams in mathematics, thereby enabling powerful insights into the materiality of mathematics itself. Châtelet, however, does not explicitly attend to the role of language in mathematics. This absence is problematic for us, because language plays a significant role in how the body comes to matter as a sociocultural entity, and it plays an equally important role in learning mathematics. Our goal in this chapter is to examine language in inclusive materialist terms, exploring how classroom discourse, in particular speech, is coupled to other materialities and affective forces. We treat speech not as a translation of thought, nor only (nor mostly) as a semiotic act of symbolic signification, but try instead to attend to the way speech and the production of sound become part of the learning assemblage in mathematics classrooms. These assemblages are continuously formed, deformed and (de)stabilized over time. In offering this material reading of speech, we hope to show how a micropolitics of classroom discourse offers new insights into the production of meaning, while de-centring the rational thinking subject as the source of meaning. This approach allows us to study ways in which language is bound or fused to other materialities, and also allows us to get at the affective forces at work in classrooms, so that we can study the way they flow across the classroom. We first discuss more generally the politics of mathematics classroom discourse and show how our approach relates to and complements this work. We then analyse a series of examples to show how mathematics classroom discourse is, in part, propelled through bodily movement and prosody.

The materiality of language

What many have referred to as the 'linguistic turn' in mathematics education research has shed considerable light on the specific ways that language operates in classrooms. In particular, recent work on the distinctive grammar of mathematics has shown how particular grammatical aspects – such as nominalization, high modality and dense noun phrases – contribute to a particular image of mathematics as objective and timeless (Morgan, 2005; O'Halloran, 2005). Studies of mathematics discourse more generally, drawing on critical discourse analysis and other discourse frameworks, have emphasized the way mathematics teaching and learning are both constrained and enabled by discursive patterns or formations found in classrooms and texts of various kinds.

Additionally, a focus on language has allowed researchers to study the political framing of school mathematics in new ways, unravelling the way that mathematical achievement and classroom participation align with socio-economic stratification and racial inequality. Walkerdine (1988; 1990), for instance, in her ground-breaking work on the production of mathematical reasoning as a cultural form, studied how the discursive patterns of elementary school mathematics came to be valued and legitimized over informal mathematical discourse. Morgan (2005) showed how mathematics texts addressed readers in particular ways, positioning students in relation to class-based assumptions about agency.[1]

This research sheds considerable light on how subjectivity, authority and agency are regulated by 'orders of discourse' and mapped onto the 'linguistic features' of texts and utterances (Fairclough, 2003, p. 29). Lemke (2002), for instance, argues that the practice of constructing intertextual relations among different kinds of language use in classrooms 'does the social work of reconstituting the social relations of the community' (p. 39), and he suggests that 'language functions ideologically not only by enabling us to make meanings that maintain the dominance of one group over another but also by not providing ready-made ways of challenging that dominance' (p. 40). These kinds of approaches to classroom discourse have led to fine-grained, turn-by-turn analysis of classroom transcripts, allowing researchers to study power relationships as provisional moves within particular language-based exchanges.

[1] See also: Atweh, Bleicher, & Cooper, 1998; Chistensen, Stentoft & Valero, 2008; de Freitas, 2008a; de Freitas & Zolkower, 2009; Herbel-Eisenmann, 2007; Herbel-Eisenmann & Wagner, 2010; Morgan, 2005; Walshaw, 2004; Zevenbergen, 2003.

This work seems particularly germane to studies of mathematics classrooms, where esoteric texts are 'recontextualized' (Bernstein, 1996), and where knowledge is *re-produced* for specific pedagogical purposes. Linguists, such as Veel (1999), O'Halloran (2005), and Christie and Martin (2007), have effectively applied systemic functional linguistics and the sociological theories of Bernstein to school mathematics texts (both verbal and written), arguing that the 'instructional discourse' is embedded in 'a discourse of social order (regulative discourse)' (Bernstein, 1996, p. 49). The technical register, the hierarchical ordering of concepts and the grammar of school mathematics constitute a variety of language ill-suited to talk of the everyday. According to Martin (2007), this monstrous 'vertical discourse' resists 'common sense' primarily because of the excessive use of nominalization by which processes and actions are reconfigured as nouns. The tension between the language of school mathematics and other language use is well-illustrated in the differences between verbal and written mathematics. Students speak mathematics with material and physical verbs – often conjugated in diverse and sometimes contradictory ways – while written mathematics shows little to no trace of these material verbs. For instance, algebraic equations (such as $20 - 3x = 46$), are formally 'translated' into dense noun phrases like: 'The difference between 20 and the product of 3 and a number is 46.' Martin (2007) suggests that the radical disjunction between the written and the spoken in mathematics classrooms, which occurs through nominalization and other linguistic practices, is possibly what makes it such a gate-keeping discipline.

Much of this literature draws on the seminal work of the linguist Michael Halliday and his extensive research on language use in context. Halliday (2007) assumes that language is a system that models or reflects culture, as well as an enactment of situational social relations. Despite the fact that the entire impetus behind Halliday's 'functional' linguistics is to shift our attention to the way that grammar evolves as a *response* to settings and situations, he continues to place the source of that response within a rational individual subject. As seen in the following quotation, he tends to characterize learning in terms of linguistic and semiotic choices from a set of resources whereby 'you build it up, and you act it out':

> What the learner has to do is to construe (that is, construct in the mind) a linguistic ***system***. That is what is meant by 'language as a system': It is language as stored up energy. It is a language, or some specific aspect of a language, like the language of science, in the form of a ***potential***, a resource

> that you draw on in reading and speaking and listening – and a resource that you use for learning with. How do you construe this potential, and how do you use it when you've got it? You build it up, and you act it out, in the form of ***text***. 'Text' refers to all the ***instances*** of language that you listen to and read. And that you produce yourself in speaking and in writing. (Halliday, 2007, p. 274, *bold and italics in original*)

To be fair, Halliday is careful to point out that the 'system' of language is not some fixed ideal structure, but rather 'the potential that lies behind all the various instances' (p. 275). Thus, the act of 'instantiation' is not the selection of one element from a finite set of choices, given that the 'meaning potential' of all texts is in fact infinite. But in our opinion he does not go far enough in recognizing the way that meaning is enmeshed in materiality, because he assumes that the relation between a situation and a text is one of 'realization', a relation which he characterizes as purely 'semiotic' (Halliday, 2007, p. 282). In separating the semiotic from the material, Halliday demotes the body to mere container or vessel for carrying linguistic resources. In other words, materiality is seen as part of the confining structure of the context, while agency is achieved only through human language use and semiotic meaning. Wortham (2006) warns that many studies of linguistic interaction are burdened by the adoption of the confining binary between structure and agency when they cast matter as part of the 'constraining structures' in opposition to the 'creative actions' of symbolic meaning (p. 279). The complex processes of material/discursive interaction, suggests Wortham, demand an approach that destabilizes the binary distinction between structure and action, and looks closely at the actions and the structures as both constraining and creative.

As discussed in Chapter 2, Karen Barad (2008) uses the term *intra-action* to speak of the relationships that bring about the distinctions between apparati and objects, and, more generally, between mattering and meaning. This term signals her move away from a concern for the way two distinct realms (words and things) *interact* and towards an understanding of what is happening between two mutually entailed folds of the same realm. Barad's move from *interaction* between two distinct realms (words and things) to *intra-action* stems from the fact that language is one of the material folds of any event. The point is not that words function within language games or simply that the words we use are constructed through sociocultural forces. This would miss the materialist and ontological argument altogether. Rather, the point is that there is some *fact to the matter* of our meaning for a word – the 'fact' is not separable from the matter. This is an attempt to grapple with the consequences of a non-representational theory of language, to grasp the gritty materiality of discourse in practice. Barad wants us to pay attention to how language or signs do not simply *correspond* or refer to the material world, but that they are indeed material configurations, and we are engaged in continuously mutating and enacting boundaries so that words and things become *provisionally* construed as mutually distinct. Language performs through material reconfiguring of the world rather than through replicating, modelling or coding a material entity outside of it. We repeat here the Barad quote from Chapter 2 to remind readers of how meaning is conceived within new materialism:

> Materiality is a discursive performance of the world, but discursive practices are not reducible to human-based actions. Matter does not serve as a mere support for discourse, nor is it merely the end product of human-based citational practices. Rather, discursive practices are specific material configurations/(re)configurings of the world through which local determinations of boundaries, properties, and meanings are differentially enacted. That is, discursive practices are ongoing agential intra-actions of the world. (p. 173)

Matter and meaning are inseparable and do not stand in a relation of exteriority to each other. One might imagine that this kind of ontology is non-operational in terms of analyzing classroom discourse, in that it seems to leave us in the lurch, without a leg on which to stand. But one of the important consequences of this relational ontology is the way it supports new research methods and new ways of studying language use in mathematical classrooms.

The challenges of breaking with such an all-pervasive binary distinction are huge, in part because language is often treated as that which definitively determines meaning, centring analyses of interaction on verbal activity. Of course, language use is powerfully implicated in meaning-making, but researchers are sometimes blinded by its apparent efficacy, unable to see how other forces modulate this power. Much is at stake in breaking with this binary, including our tendency to imagine the speaking subject as the emblem of political action. One means of pursuing a micropolitics of discourse is to study language use in ways that resist the affirmation of a human subject as a self-governing legislator or self-authoring subject, affirming instead the 'positivity of an opaquely becoming subject' (Colebrook, 2008). This sort of becoming is an 'accident-charged relational process', a form of experimentation that does not begin or end in the enthronement of an episto-empirical subject. Because learning can and often does entail a creative encountering (with norms, bodies, concepts), and because such encountering entails cutting and reassembling relations *in new ways*, we are keen to study language as in(ter)vention rather than communication. According to Colebrook, this need to rethink the nature of *becoming* is why Deleuze and Guattari suggest that affect be studied as the genetic engine of *new* systems rather than only studying *power* as the controlling engine of *current* systems. We follow this approach in this chapter so that we might show how mathematical activity emerges in and through a 'body' that exists in 'relations, connections, creations and syntheses that are distinct from any organized body' (Colebrook, 2008, p. 127). Micropolitics is offered as a way of studying the emergence of political structures or assemblages produced through forces that operate alongside the political agent.

In order to rethink language and matter together, and to pursue a post-humanist account of their entanglement, it can be helpful to consider a vast, geologic perspective where language and matter evolve as one entanglement (Delanda, 2008). Language, according to this grand narrative, emerges biologically as a mode of expressivity, unfolding (and extending) in both temporal and spatial relations. Thus, language is merely another kind of 'expression' – a particular arrangement of material relations – like 'colour, sound, texture, movement, geometrical form and other qualities that can make geological or meteorological entities so dramatically expressive' (p. 163). We find this geologic perspective provocative in that it demands that we study language use outside of a theory of communication and representation. Instead of seeing language as the transcendental coding of all meaning, we suggest studying it as a material expression so that we might begin to grasp the material work that it does. By inserting the parentheses in the word in(ter)vention, the term takes on a dual meaning, highlighting the way that language operates on at least two planes, intervening in and reconfiguring current learning assemblages, and simultaneously bringing forth or inventing new materialities through discursive 'cuts' and diagram-gestures. The term in(ter)vention captures the sculpting and ontogenetic nature of language. Thus we hope to study language as *part of* the material assemblage, rather than as representing or coding it. The challenge is to examine the material coupling of speech and meaning, in order that one can study how language effectuates and is itself effectuated: To 'order, question, promise, or affirm is not to inform someone about a command, doubt, engagement, or assertion but to effectuate these specific, immanent, and necessarily implicit acts' (Deleuze & Guattari, 1987, p. 77). In other words, speech and other language acts have to be studied for how they couple with other materialities and operate outside of a communication model.

Word assemblage

The communication model of speech tends to insist on the existence, intention and rationality of an interior life that transmits information between private and public spheres (Massumi, 2002). This model all too often treats speech as simply a matter of representing, describing, corresponding to or complying with prior content or information, as though our spoken words were nothing but the carriers of otherwise disembodied meanings.[2] If we examine this model carefully, we find a deep-seated conviction about the ontological distinction between meaning and matter. This conviction longs for speech to be the linear causal link between content and expression, and thus reflects in part our own anxiety around non-sense and a world inflected by chance. What if the potential of speech were linked to non-sense, accident or singularity? Here the term *singularity* is not meant as an instance of a universal, but as an irregular occurrence or disruption. What if the potential of speech was in its capacity to disrupt and nonetheless act as a magnet in structuring the behaviour of others around it, much like a singularity does in a differential system? Rather than conceiving of speech acts as instantiations of a linguistic system, we want to study them as potentially singular acts of in(ter)vention.

Following this line of research allows us to study how speech is actually sutured to the situation, and yet the stitching itself is strangely ungoverned. Something in speech erupts like a spark or singularity, migrating across the material network and recoupling with gestures, objects and patterns, so that mattering and meaning are conjoined processes. The force of speech is in this play of difference, where pre-articulate intermediate entities flit about without resting easily on either side of a fledgling boundary between meaning and matter. As much as we polish and hone our language use and are subject to code and systemic meaning, speech seems to sustain this indeterminacy in its stuttering fumbling ways. Attending to this indeterminacy and underside of speech demands that we think of word-sounds as both symbolic of meaning and 'asignifying particles' of expression. In educational research, we comb over transcripts for evidence of mastery and misconception, and we look for how students and teachers use words to carry out identity work, but we also need to look at how speech is operating outside of a regime of signification – outside or alongside its capacity to refer, and even to communicate. We need to also look at speech as part

[2] Sfard (2008) is an exception; indeed, in order to signal her distance from this point of view, she coins the word 'commognition', which is meant to underline the way in which thinking *is* communicating.

of a material expression of asignifying particles, frequently atypical and stammering, and often constituting the 'cutting edge of the deterritorialization of language' (Deleuze & Guattari, 1987, p. 99).

We suspect that speech often falls through the 'propositional mesh' of signification, evidencing a potential force or flow that cannot be captured (Massumi, 2002, p. 24). The sound of mathematics problem solving offers a case in point, as it rarely, if ever, unfolds in a smooth, linear emplotment of solving. This is made evident by Rowland (1995), in his study of hedges, such as 'no, well, sort of ... maybe' and other traces of epistemic and ontic modality ('sometimes, always, maybe, certainly, supposed to, might, have to, want to') in the speech of children as they worked on mathematics problems. In the following transcript, one can see the incredible expansive and non-linear movement of a boy's speech as he explores the following problem: Given any square array of dots, what is the minimum number of lines you need to connect all the dots? As you read, imagine how his tongue touches the place in his mouth where the sound of the potential conjecture is tentatively formed, how he prevaricates with the requisite modulation but also creates massive amounts of space where his words can make contact with the various parts of the problem. We can hear in this transcript the way that speech has its own aleatory force, which is often at odds with the task of communication:

Teacher: All right then, so what are you going to do now?

Student: I'll try a, um, four by four grid.

Teacher: Right. Can you make any predictions before you start?

After some hesitation, the student eventually replies:

Student: The maximum will probably be, er, the least'll probably be about 15.

Teacher: So why did you predict 15?

Student: Uh ... because I thought there might be a pattern between ... if there was um, a certain amount of, um ... if it's three by three say ...

Teacher: Uh-hum.

Student: If you ti ... three times three is actually nine.

Teacher: Uh-hum.

Student: But as, if you went round all the dots, it would only come to about, if you did it once it would come to one, uh, less than nine, you got, uh, because, because there's o ... there's only ... cause you only have, y ... you can miss out a line exactly, cause you, you can miss out a gap, cause you, um, y'd have to go all the way round the whole dots.

> *Teacher:* Ok ... so why did that make you say 15?
>
> *Student:* Because, uh, f ... for the same reason, cause if you w ... tried to go round the whole all the dots you'd get 16 but if you just did it once all the way round the dots but missing out gaps you'd still come to uh, you just minus one basically and just ...
>
> *Teacher:* So what would happen in some other squares?
>
> *Student:* Probably if you minus one from the ... if you square the number you'd probably find that if it was actually, if you minus one from that you'd probably find that that would be the answer ... (Rowland, 2000, p. 185)

When the student is repeatedly asked why he predicted 15, his responses seem to stumble over the logical phrasing of inference – he stammers his way through the use of 'because' and the appeal to reason and if-then conditional language. But rather than merely interpreting this as a reflection of the boy's hesitation and uncertainty, we suggest that these sound-words are evidence of the prosodic facets of speech – the use of rhythm, stress and intonation – that allow the speaker to mouth the words and deform their contours, to invest or divest them of intensity, to stretch them out over the surface of the problem. This is how speech hooks up with the task at hand – the student slowly but surely sutures together his spoken words with the array of dots and other sensations, including the unfamiliar taste of the language of explanation. Rather than decoding his words in terms of how they communicate his uncertainty, we challenge the reader to imagine these words as a means of linking up his body with the body of the problem. Indeed, if his word-sounds failed to function in this way, we suspect that speech would not do the job it has to do. Rather than interpret his stumbling as a deficit – where the goal might be to communicate with perfect clarity – we see these points as suture points where his speech latches onto the problem, and his utterances pursue a new rhythm or new temporality that is part of the reconfiguring.

This transcript begins to show how speech operates entirely differently from writing. If writing is taken to refer to all the scribbling and graphical activity that is so common to mathematical practice, functioning to some extent as 'the symbolic ecology of mathematical thought' (Rotman, 2008, p. 1), then speech plays a highly unusual role in teaching and learning mathematics. Classrooms that focus on communication often treat speech as though it were merely a translation of completed thoughts. In the literature on communication in mathematics education, one often finds a 'replication' model where speech comes after thought, and students are meant to share their prior thoughts in spoken utterances

(Kastberg, Norton & Klerlein, 2009).[3] This replication model can be found in common practices that request students to report on their prior thinking, in phrases such as 'put your thoughts into words' or 'explain your thinking'. And although we believe that these practices have significantly increased participation in mathematics classrooms (and should be commended for that reason), our objective here is to unpack these practices for how they operate through tacit assumptions about thought and its relation to language. Restating and revoicing have also become common teacher practices in mathematics classrooms, often entailing a revising of a student's utterance in technical mathematical terms. Indeed, 'think aloud' strategies, whereby teachers narrate their own thinking while they solve a problem or investigate a conjecture, are also used to 'give voice' to mathematics. But the practice of 'think aloud' is often premised on a translation model in which speech is a copy of thought. Bochicchio et al. (2009), for instance, posit five essential components of successful student communication, the second being 'publicizing' student thinking. The translation model carries with it certain assumptions about language use, often treating speech as mere lip-service (to prior and completed thoughts), rather than treating it as the actual site of thinking and unscripted in(ter)vention.

We can see this translation model at work in many classrooms. Consider the video of a grade seven mathematics classroom in California, which shows a lesson that aims to develop student algebraic thinking through generalizing arithmetic tasks.[4] What strikes us immediately when watching the videos is the considerable emphasis on oral communication and whole-class conversation. Although the camera records some of the student contributions during small-group work, the focus is primarily on the whole-class segments. Students' hands shoot up, as many want to contribute. As the authors state in the introduction, 'Student thinking and sense making are at the heart of each case, as we see students wrestle with new ideas and conceptions of themselves, of mathematics, and ultimately, of their world' (Boaler & Humphreys, 2005, p. 4). The teacher adeptly calls on students, 're-voices' their responses and asks other students to re-voice, as well, while she simultaneously documents on the white board their various procedures. Students are invited to go up to the overhead projector and use the diagram as they explain. From the perspective of the reform

[3] An exception to this can be found in Stockero and van Zoest (2011), who suggest that teachers need to move beyond this 'replication' model.

[4] This video can be found in a teacher education resource entitled *Connecting mathematical ideas: Middle school video cases to support teaching and learning* (Boaler & Humphreys, 2005).

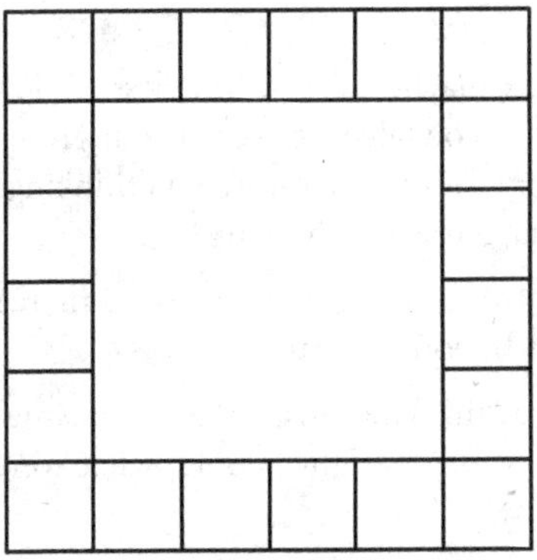

Figure 5.1. The image projected on the wall of the classroom.

mathematics education movement, this focus on communication and multiple representations is good pedagogy. Indeed, the teacher does an impressive job of moving the students towards generalising their procedures, so that they might begin to think 'algebraically' about finding the answer to a more general question. But it is precisely this emphasis on the verbalizing of thought that seems problematic in its implementation, despite the fact that 'thinking aloud' is considered good practice in mathematics classrooms. When the video begins, there is a diagram of a 10x10 square border (picture frame) on the overhead projector (Figure 5.1), and the students are sitting in groups at the their desks. The teacher asks the students to determine 'mentally' how many unit squares are in the border.

The teacher asks them to do so 'without talking, without writing, and without counting one by one'. Even though many students have their hands raised, the teacher asks them to 'talk to each other just about what you got'. After some discussion, the teacher calls attention back to her and asks, 'who has, by the way, do we all want to say the answer?' At this point, the students collectively chime, '36. The teacher asks whether some students had originally thought the answer was 38 or 40, and then asks those students to explain 'what were you thinking?' Then, several students are asked to describe how they obtained their (correct) answer. As different students speak, they recount their different methods, clearly familiar with this classroom practice of 'verbalizing their thinking'. After listening to the video numerous times and counting and mapping the way this utterance ('what were you thinking?') occurs, we began to hear the refrain as a command, repeated again and again as the teacher moves from student to student. One begins to notice the powerful effect of 'what were you thinking?' in how it structures the space of possible utterances

What are you thinking?

1	*Teacher:*	Did anyone think it was maybe thirty-eight first? A few people, OK. And, so, when you talked it over in your group, those of you [who] thought forty, what were you thinking? All right, Stephanie, what were you thinking?
2	*Stephanie:*	Uh, I was thinking that one side is ten and then there's four sides and times ten by four is forty.
3	*Teacher:*	OK. How many were thinking just like Stephanie? All right, what about the thirty-eight people, what were you thinking? Uh, Mindy?
4	*Mindy:*	I was thinking about the top two are ten and so ten plus ten equals twenty, and then the other two to get were nine each, so that really makes eighteen, and twenty plus eighteen …
5	*Teacher:*	OK. All right. So now let's just see some different methods – you know how we like to gather different methods. Let's see some different methods for getting the thirty-six. So, let's see, let's have Sharmeen. Sharmeen, what's your method?
6	*Sharmeen:*	Well, Krysta started out with forty from ten times forty [four] and then I subtracted four from that because there would be four squares overlapping, and so that got thirty-six.
7	*Teacher:*	Raise your hand if you understand how Sharmeen did it. And the four was for
8	*Sharmeen:*	The squares that overlap.
9	*Teacher:*	OK. And this four was for
10	*Sharmeen:*	The four sides.
11	*Teacher:*	And this ten was for
12	*Sharmeen:*	The ten squares on each side.

while simultaneously *performing* what it means to think. In the transcript excerpt and elsewhere in the video, Sharmeen uses language so obediently and masterfully that she is often called on to manage the students who stutter more. Her speech is obedient (and masterful) in that she says exactly what we suspect she knows the teacher wants to hear, with a smooth delivery, no hesitations or repetitions, and an unwaveringly linear narrative structure that uses the temporal connective 'then', the past tense ('I did this, and then this and then that') and inferential grammatical terms ('because', 'and so').

But there is an important moment – a singularity – when the student Colin is called on. Up until this moment, students have stayed seated while responding, but Colin's contribution functions as a singular perturbation that redirects the classroom interaction. Colin's problem-solving method – framed, like the others, in the past tense – involved starting on the right edge of the border. He first saw that the right edge was 10, and then he moved to the adjoining side and noticed that there would be 9 remaining 'because this one's overlapping'. Then he moved to the left one and noticed that it would have 9 for the same reason. Finally, he looked at the last edge and noticed that it would have only 8. In the classroom interaction, however, we see that his language use is less a matter of communication and more a matter of making material assemblages with the diagram and any other devices he can find. Perhaps most importantly, the teacher *effects* these new assemblages through her own verbal intervention.

In his new material surroundings, Colin uses a series of indexical expressions ('one side', 'bottom', 'it's ten', 'that was') that bind his speech to the actual diagram on the overhead projector. This indexical language allows his speech to engage materially with the diagram, rather than only functioning as the vehicle of explanation and truth claims. Use of indexical language ('here', 'there', 'this') conjoins the speaker's body with the diagram and invites others to engage with the problem, as it is located on a workable surface and in the present moment. Instead of a language of naming that detaches the mathematical entities from the visual plane, indexical language is haptic, quite literally touching and handling and situating the entities on the plane of public inscription. Colin is not simply using language in an indexical way *to point at* material inscriptions, for this would be to demote speech to reference. Language use is not the legislator of meaning here, but is a new force when introduced, literally adding more matter to the ever-altering material assemblage. When the teacher asks that Colin go up and show the others, he confidently asks if he can use the teacher's 'pointer-thing' to engage the diagram. In doing so, a new material assemblage is formed (Colin-'pointer-thing'-diagram), where language becomes materially bound to these other elements, rather than only functioning as communication or representation. The teacher's invitation thus occasions this new alteration of the material assemblage, which includes both the teacher and Colin's peers. Indeed, the event draws the watching eyes of Colin's peers to his hand movement (on the glass of the projector) and the shadow of this movement projected onto the screen on the wall. Colin produces a moving image for his peers to study, gesturing the diagram into activity. This doubling of his body

A new material assemblage

13	*Teacher:*	OK. Another way to do it? Colin?
14	*Colin:*	All right, how I did it was ... I just put one side was obviously gonna be ten ... so it's ten, and then I did like the bottom one and that was gonna ...
15	*Teacher:*	Could you go up and show us? I think that might help us.
16	*Colin:*	Can I use like a pointer-thing?
17	*Teacher:*	The pointer's in the top drawer if you want to use it ... top ... right in the center ... there it is.
18	*Colin:*	All right I found it. Anyway, so I, like, I know this side is ten so I just did like ten and then this one, this one's the overlapping one, so then this would be nine, then this one would be nine, too, because this one's overlapping. And then for this one it'll be eight because these two, this one is being used by this one and this one is being used by that one.

(his hands at the projector, and the shadow of his hands on the screen) produces another fold in the growing assemblage, another dimension in the multiplicity of his ever-changing embodiment. Colin's body is a moving, quivering, unstable collective; his speech and his gestures are bound up and sustained in this collective. In line #18, Colin embeds his speech in the diagram to such a degree that his spoken words become material appendages of this new assemblage. The repeated use of indexical language further decentres language as the legislator of truth (and the vehicle of explanation) and invests such power instead in the material 'pointer-thing', a device that Colin manoeuvres – testing his body, his reach – with great joy. This moment in the video marks a shift in the classroom activity, when Colin moves across the room and interacts with the diagram, thereby breaking up the stasis, and altering the flow of energy across the learning assemblage.

We want to examine this moment in more depth, as it shows well the materiality of discursive practices. In the following rearranged transcript, the lines have been coupled into speech/object/gesture apparatuses that are reconfigured from moment to moment. One notices that sometimes the object/gesture begins before the indexical word is uttered, while at other times the movement follows the sound-word. Indeed, the challenge lies in studying the complex simultaneity of all this activity; at the micro-scale, we begin to glimpse the pre-perceptual forces that seem to disrupt and disperse any simplistic reading of this activity as *merely* communication. Of course, the transcript still uses words to describe the event, but rather than insert stills from the video – which would also open up new dimensions – our aim here is to show how one might create a transcript that uses language to decentre language. We have tried to show how the utterances are stretched out across other material activity, and how the duration of the speech is rhythmically bound up with other bodily engagements. We have bracketed the spoken words rather than the acts, and have thereby playfully overturned the usual practice of transcription. We have also tried to highlight the fact that the teacher and students perceive Colin as he manoeuvres the pointer-thing at the projector, while simultaneously perceiving the image of the pointer-thing as it moves across the projected image of the diagram. If we dwell on the former, we ascribe material agency to Colin, but if we focus on the image, we ascribe material agency to the moving pointer-thing. Even Colin, in his own use of language, verbally ascribes agency to the sides of the border, suggesting that certain edges 'use' certain squares. Thus, the situation offers an excellent example of how our perceptual capacities perturb any easy assignment of agency to a sole human agent.

pointer emerges from drawer [anyway, so I like]
shadow of a hand appears at the edge of the projected diagram [um I know]
pointer moves down along right edge [this side's ten]
Colin tilts head towards diagram, hand rests, pointer rests [and so I]
movement of the pointer along right edge with increased speed [did like ten]
Colin's hand gestures generously along the edge, the gesture extending beyond the corner
shadow of Colin's hand crosses the diagram, pointer taps on bottom left corner [and then this one]
pointer moves slowly to right corner, taps bottom right corner [this one's]
[the overlapping one] sound of pointer tapping the glass of the projector [so then]
pointer moves along bottom quickly towards left corner, [this one would be nine]
hand moves right along bottom edge [then this one]
hand moves along left edge, resting at bottom left corner [would be nine too]
bouncing pointer on bottom left corner [because]
Colin's wrist twists so as to spin pointer-thing around at the left corner of diagram [this one's]

[overlapping] distant sound of chalk scratching on the blackboard [and then but]
body and image still [it'll be eight]
pointer makes a non-linear snake move to touch each corner of top edge [because] [these two]
[this one] tapping left edge [is being used by this one] tapping right corner
sound of chalk on the blackboard [and this one]
tap-tap [is being used by this one]

We can now see in this alternative transcript how the movement of the pointer gives meaning to the inferential words 'then' and 'because'. It is not enough to say that Colin's gesture communicates information about what this 'because' means, or that the pointer's movement expresses spatially and effectively something that is too vague, or awkward, to express in words. Both these interpretations fail to do justice to the materiality of the spoken word, the moving pointer and the projected image, and to how they are assembled together. The 'because' is part of the same flow of intensity as the pointing to the corner square, which has already been touched twice by the pointer's motion along the bottom and left edges (and thus counted twice). The meaning of the word 'because' comes from the material configuration of the 'this' edge and the 'one' unit squares that Colin has extracted from the original diagram, which creates a local determination of which squares can be 'overlapping' (although no such overlap is actually visible). The very use of the word 'overlapping' insists on the materiality of these counting gestures. Far from being only a simple, logical connector, 'because' identifies the corner square as one which has multiple functions – as participating in two different ensembles. These terms of inference are part of a material reconfiguring of the diagram, part of a boundary-making practice – to use Barad's phrase – that shows how meaning is bound up with the perceptual contingencies of the assemblage. But the movement of the pointer acts with equal force to the words that are spoken, co-generating the meaning of the words and investing the explanation with validity.

In the case of Colin, we see how his words reassemble the word-sound-diagram-pointer links, and how the sounds he makes can be seen as asignifying particles or intermediate entities that, together with the hands, the projector, the diagram and various other material-affective forces, produce an entirely new mathematical assemblage. Perhaps speech functions most effectively in mathematics classrooms when it joins with other materialities in this kind of way. Although linguistically inspired approaches to mathematics (classroom) discourse help us trace the grammatical patterns that legitimize participation in the field (Whitin & Whitin, 2002), they all too often impose a regime of signification on interaction, explaining away the accidents at work in language as mistakes or misconceptions. Different kinds of approaches to language use may help us study the way that speaking mathematics entails a far messier world of becoming mathematics. Thus, the refrain 'what were you thinking?' might actually be enforcing the use of language as a recording device rather than a creative medium. Placing emphasis on verbal explanation, despite its excellent objective to make mathematics more public, can often seem like submission to an 'aura of signification' (Lecercle, 2002, p. 249). In such classrooms, it can often feel as though the students are burked by language, as though 'a wet blanket of signification smothers any attempt at expression' (p. 6). If thinking is in the flesh or is somehow distributed across an assemblage in which language participates, then communication as replication of thinking fails to address the complexity of learning. Perhaps the philosopher Michel Serres (2008) is right when he tries to decentre speaking and attend more vigorously to the way in which touch and other sensory modalities function in our lives; he claims that 'words fill our flesh and anaesthetize it' and that speaking can function like a 'discursive breastplate', a shield to numb our encounters (p. 59). Similarly, Hwang and Roth (2011), in a further effort to centre the body as the locus of learning, broaden their conception of communication to include all sensory modalities – gesture, intonation, appearance, smell, etc. – showing how 'thought is dynamically related to the whole unit of communication rather than to words alone' (p. 28). They argue that 'the living body constitutes the mediating hub in communication; my body is my expression rather than merely a tool for expressing what is in my mind' (p. 29).

The human voice

For the ancient Greeks, hearing was the preferred sense, in large part because of the perceived immateriality of sound, which linked it to a pure, ideal realm (Ong, 1982). Thus, for Aristotle, the sense of hearing had a special connection to notions of character and morality. The high status of hearing also privileged certain art forms, like singing and oral poetry. The human voice was given a privileged place: 'The human voice was the most perfect instrument for giving expression to the inwardness of the soul and thus accomplishing an "objectification of subjectivity"' (Rée, 1999, p. 60). Voice seems to come from the inside, to emerge untainted from the interior. But what it produces – sound – never lasts and never asserts a precise fidelity to the 'outside' world. Rée suggests that:

> [T]he fundamental difference between auditory and optical mistakes arises from the fact that we are less tempted to attribute 'outness' to what we hear than to what we see. Hearing does not presume as much as vision. It is not so arrogant, and it is willing to refer its experience to evanescent qualities without insisting, as sight does, that they have to be tethered unambiguously to definite things in the material world. (p. 46)

Indeed, it was the evanescent and seemingly immaterial nature of speech that led both the ancient Greeks and various philosophers to assert its primary status. Ingold (2007) has argued that, for ancient Greeks, the word and its sound were seen as a unity, so that, for example, the words of a song carried with them that song's melody. The meaning of the word was essentially tied to its melodic sound. The modern separation of words from musical notes has broken down that unity, so that voice now has a double nature as producer of meaning and producer of sounds.

In his book *I see a voice*, Rée (1999) closely examines the nature of the human senses and the implications of their particular features in the development of cultural forms. He opens with the statement that '[t]here is nothing more personal than voice' (p. 1), and he argues that the history of the senses entailed the slow but sure loss of human voice in favour of written language, whose main sensory access is the eye. This is a familiar historical claim, one proposed by Ong in his seminal work *Orality and Literacy: The Technologizing of the Word* (1982) and thoroughly critiqued by Jacques Derrida for its Western propensity to cherish and centre the human voice and embodied presence. Rée, however, alters the account somewhat, suggesting that voice exhibits 'an intriguing human tension, even a contradiction: [It is] both expression and communication, both feeling and intellect, both body and mind, both nature and culture' (p. 16). He is at pains to remind the reader of the intimacy of voice, its textures, pitch, rhythms, volume and dynamics, as well as its complex kinaesthetic and oral production. Why do we forget that the voice is so personal, so furtive, so carnal? What would it mean for us to attend carefully to the materiality of the voice? To the muscular contortions of the lungs, throat and mouth; to the sounds that are less sense than non-sense, the producing of gurgles, sighs, stammers? With the emergence of writing, it was precisely these gurgles, sighs and stammers that for some made the human voice suspect and dangerously open to irrational forces.

As writing became more thoroughly systematized with the printing press, speech remained the main medium for the 'chanting of spells, the signing of charms, the swearing of oaths, and all kinds of abracadabra, mumbo-jumbo and verbal hocus-pocus' (Rée, 1999, p. 3). It was this expressive voice – its dark, moody and temporal substance – that seemed to lack credibility and validity. Rée sees Descartes as wanting us to 'break out of the ancient prison-house of speech and one-dimensional temporality, and disport ourselves in the multi-dimensional spaces of writing or textual production' (pp. 6–7).[5] For Descartes, speech is trapped in the moment (and the mouth), and human rationality must transcend these specificities. While speech still functions centrally in contemporary contexts such as law courts and talk radio, and there remains a tendency to trust the verbatim or that which was 'actually said', Rée argues that modern life centres rationality on the written word, leading to a 'victory of an ocular civilisation focused on an impersonal external world' (p. 4). This ocularity has been achieved not so much through a privileging of vision, but through a silencing of the human voice through the technology of alphanumeric writing.

Alphabetic writing seems to wrestle speech away from the body, to erase the traces of any corporeal dimension, operating as a system of symbolic notation ultimately detached from the organs that produce and perceive the sounds. In Chapter 8, we discuss in some detail how Châtelet studies mathematical notation for the ways in which writing actually retains these traces, and indeed, most of this book is concerned with showing how the various inscriptions of mathematics entail material encounters between bodies. In this section, however, we focus on the distinctive relationships of which speech partakes. In particular, writing fails to evoke the prosodic effects of speech, the ways in which an utterance is forceful precisely through its musicality. These are the gestures within speech that lay claim to a multi-dimensional space – 'the tone, the rhythm, the variation of emphasis, the loudness, the changes of pitch, the mode of attack, discontinuities, repetitions, gaps and elisions, and the never absent play of musicality of utterance that makes human song possible' (Rotman, 2008, p. 3). Speech seems to operate through an entirely different materiality, a different set of potentialities and a different relation to the flesh of the body, and these differences are evidence that speech has its own creative material

[5] Perhaps a contrasting position is that of Kant, for whom neither vision nor hearing was as strongly associated with civilised life as speech; he went as far as to assert that the mute could never attain the faculty of reason.

force. In other words, speech augments the learning assemblage by literally adding more material, projecting and binding with other matter, and linking up with other nearby frequencies in distinctive ways.

An interesting example can be found in Greiffenhagen (in press); it shows how college mathematics lecturers mark the writing surface of the blackboard in particularly rhythmic fashion, so that the production of the writing is syncopated with their spoken words. In the following excerpt (see Figure 5.2), a lecturer puts the first line of a proof at the top of the blackboard and the last line of the proof at the bottom. As he makes these two inscriptions, he uses his entire body, together with the spatial metaphor of 'somewhere', to map the desired trajectory of a proof by contradiction. It is not that the proof must literally be contained within the given allotment of space, but that the listener should conceive of the spatial and temporal wholeness of the proof. One can see in this structural or schematic act of inscription an attempt to frame the activity in terms of space, time and logic. The lecturer marks the contra-positives of premise and conclusion so that the listeners will not suffer any surprises or meander about without determinate goal – that is, so that the listeners will not be distracted by any tangential spoken words and so that the flow of deduction will thread its way through all that is said and written. It is through the human voice that the students learn about the speed of the proof – the rhythm and tempo of its deductive steps. This point is worth elaborating slightly, as it is often ignored. When presented with a written proposition

So. ((erases theorem))
((starts writing at top))
What we've done here, is assumed.
It's not-, it's not possible to prove
theta from Gamma. And somewhere
at the bottom of the page.
((starts writing at bottom))
We want to get down to showing:
Therefore gamma does not logically
imply theta
((finishes writing))

Figure 5.2. A mathematics lecturer's blackboard.
(*Source*: Greiffenhagen, in press)

or complete proof, our eyes are able to move quickly across the visual notation at almost any speed, depending on our familiarity with the truth of each proposition.

$$\begin{array}{l} p \rightarrow q \\ q \rightarrow r \\ p \\ \therefore \quad r \end{array}$$

If we think of inference as an event, it is an event that fails to convey or stipulate the speed by which it occurs, and hence many mathematicians and philosophers consider logical inference as atemporal. In this book, we pursue a materialist reading of mathematics, primarily focusing on how mathematics is infused with mobility and movement, but temporality operates through diverse means, and one of those is the human voice. The lecturer, for instance, uses the human voice to attend to the slowness of certain deductions and to help the students partake of the different speeds of the proof. The 'steps' of every proof have different speeds in terms of what is both formally and materially entailed in the particular step. Although we often interpret this in terms of difficulty, such an interpretation demands further analysis. What makes it more difficult? Lecturers frequently frame their problems and proofs in terms of the anticipated difficulty or ease of the task at hand – 'this is a hard one' or 'you'll probably find this part easier' – but we think it is worth unpacking this construal of difficulty in terms of temporality and duration. We suggest that the lecturer's ongoing speech throughout the proof functions to modulate the changing speed of the proof. In doing so, the lecturer's spoken words are coupled to the necessity of logical inference, and thus necessity and contingency co-inhabit the proof.

The entire tradition of the blackboard lecture – which is so dominant in tertiary mathematics – pivots around this complex entanglement of speech and writing. The lecturer's speech cannot simply be treated as a limited representation of the seemingly more concise writing, for speech has its own micro-gestures of intonation and stress. The binding of sound and meaning in the classroom has to be examined at this micro-level and analysed for how affect and rhythm, together with other sonorous facets of sound, produce meaning. Operating alongside the visual gestures are the gestures of speech, or what is called *prosody*. Attending to the prosody of speech allows us to focus on how the movements of the body are essential in making speech – how talking involves the curling of a tongue and various minute vibratory actions of the face and body. Rotman (2008) cites the evolutionary neurologist Terrence Deacon, who argues that these movements are integral to the making of meaning, in that one listens to the movements (minute vibratations) as one makes sense of the spoken words. The listener focuses on the preparatory movements, like the pauses, accelerations, fallings away and other bodily performances that produce the sounds, rather than merely attending to the discrete sonic units or speech sounds. These apparently paralinguistic elements of speech production figure far more prominently than linguistic analysis of classroom discourse would have us believe.

The prosody of the spoken utterance is the material link between meaning and matter, operating as a kind of indexical coupling between symbolic reference and the gestural indeterminacy of the body. Given that mathematics is a practice steeped in ideograms and diagrams – a practice whose meanings are encoded in written text – the presence of prosody in speech (and its absence in writing) is hugely relevant to the making of mathematical meaning. How might prosody in the classroom function in making sense of mathematics? How might speech in the mathematics classroom be different precisely because of the way prosody is enacted? Roth (2011) has begun to address these questions in his analyses of a grade two geometry classroom. In particular, he attends to the way in which rhythm 'structures the ways in which we communicate' (p. 160) in terms of, for example, the rhythmic ritual of turn-taking in a conversation. When analyzing classroom conversation, Roth measures the pitch levels of speakers and shows how they align when there is agreement or solidarity among the students and the teacher, how they move higher and higher in situations of conflict, and how students reproduce pitch contours of a teacher's utterance. Roth illustrates the last phenomena using a time/pitch graph that shows turn changes. Similarly, we could show the almost perfect pitch matching in the transition between turns 17 and 18, where the Teacher and Colin are talking about the 'pointer-thing' (see Figure 5.3; the vertical lines show the turn changes).

Such graphs provide strong evidence for Roth's claim that rhythmic phenomena make interaction possible. Drawing on Nancy's (2007) affirmation that 'sound and sense mix together and resonate in each other, or through each other' (p. 7), Roth also sees sound rhythm as underlying the very constitution of sense. Roth's attention to the rhythmic nature of conversation in terms of resonance, however, reflects his focus on consensus-making in the classroom. In addition to looking for resonance, it is also important to study the micro-level of pitch and modulation in terms of dissonance, when creative lines of flight and divergent thinking are carried forth in what one might call contrapuntal harmonies.

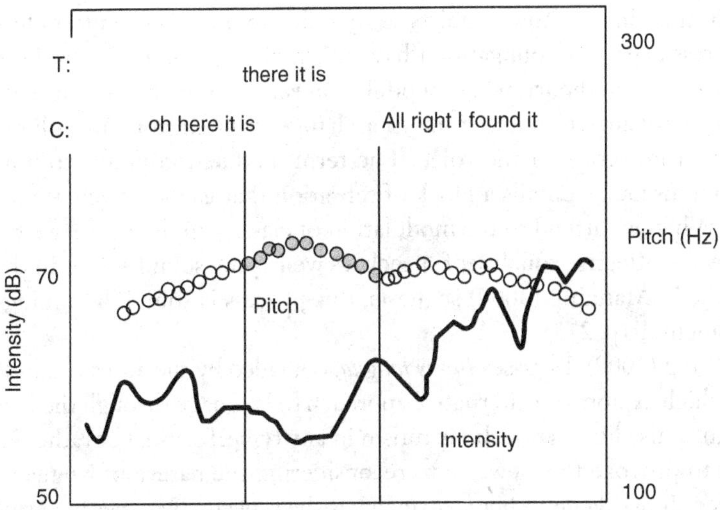

Figure 5.3. Time/pitch graph showing pitch matching.

As we discuss in Chapter 7, the generative power of non-sense needs to be studied more carefully in mathematical activity, where non-sense nevertheless operates through the senses – often as that which is taken to be imperceptible – as well as through paradox and mathematical monsters. In the context of this section, the issue of non-sense relates to how we decide whether or not some sounds are noise. Roth might be ignoring the constitution of sense within these other sounds, seeing as these are emergent properties of speech that are not decoded in terms of communication content. Although it is possible to encode aspects of prosody in transcripts, as some researchers have done (see Jefferson, 2004), we are interested in how the prosodic and the syntactic seem to follow different rhythms – which are not always resonant – and in how our analysis of classroom discourse must reckon with its polyphonic and dissonant nature. Indeed, one of the challenges for work on speech recognition technology is to identify a stable unit of pitch, accent or stress. In other words, prosody is an unfolding event that resists being packaged into stable units. One could argue, and many have argued, that the power of the alphanumeric, and perhaps of writing technologies more generally, is in its system of discrete units. The prosodic dimension of speech, on the other hand, is a continuously changing quality or differential that will forever outstrip any attempt to be unitized. For this reason, the term *modulation* (rather than mediation) seems well-suited to the study of classroom discourse. We discussed earlier in this chapter how linguists study modality in terms of utterances that express degrees of obligation ('have to', 'must', 'would like'), but here we draw on music theory, where modulation refers to the transition from one key or tonality to another, or to a change in pitch, tone or volume through an inflection of the voice. The term modulation points to how classroom discourse entails a block of sensation that is collectively intensifying. When we attend to the modulation of classroom discourse, we try to follow the affective tonality of speech, as well as the sound-words in the utterance. As Manning (2009) suggests, this process is sound 'becoming-environmental' (p. 213).

Manning (2009) discusses *In my language*, a video by the autist Amanda Baggs, which explores an alternative approach to language through the lens of autism. This video is not about autism in any typical sense, but rather an attempt to provoke the viewer into reconsidering the nature of language. The video shows Baggs in her apartment, as she repeatedly moves her arms up and down or rhythmically strokes a doorknob or presses her face into the pages of a book. The sounds of each of these haptic encounters play a crucial part in the sensory experiences of the non-verbal autist Baggs,[6] whose typed 'language' has been 'translated' into a digitized voice that is also part of the soundtrack.[7] For Baggs, dwelling on the felt encounters with material objects can be considered a kind of communication.

> For Baggs, communication through words remains inadequate to the singular experiences of sensation the world calls forth. Words are an extra component of the experience of articulation, not its final form. Words, Baggs suggests, cannot fully express experience's complexity. Her 'translation' must therefore evoke more than the manifestation of words. It must transduce the event of language's becoming-with-sensation. Language must be called forth as a layering-with of the affective tonality of expression. (p. 215)

The video provokes the viewer to consider how language and communication are linked to affective resonance and movement. We can begin to imagine a world where meaning is replete with tactile, visual and sonorous dimensions, where we feel the text come alive with affective tonality. The world is *not* simply waiting for us to impose meaning on it – there is an intensity of energy flowing through the space of encounters, enveloping us and sustaining our participation. Baggs's video humbles us, demanding that

[6] We are conscious of the issues around use of the term 'non-verbal' in relation to autism and use the term here to refer to how Baggs is not speaking in the video.

[7] This video is available on youtube at http://www.youtube.com/watch?v=JnylM1hI2jc.

we appreciate how bodily encounters are not just examples of a different kind of signification, but instead (or in addition) they are direct material encounters and entanglements with other materialities. In other words, the term 'body language' has to be troubled so that we do not simply treat bodily activity in our familiar linguistic terms. What would it mean to study verbal language use primarily as affective tonality? How can we centre the movement of affect across a learning assemblage? Of course, researchers of classroom interaction face incredible difficulties in studying affective tonality because we have been trained to decode communication in terms of learning goals or what many would consider the appropriate use of visual representations. Researchers are not meant to *feel* that which they study, and the point that Manning makes with Baggs video is that observation, including that of researchers, entails micro-perceptual and 'prehensive' forms of engagement that are indeed forms of material entanglement. One advantage of the term 'affective tonality' is that it underscores the fluid, rhythmic and collective nature of emotion or affect in the classroom.

Let us return to the classroom video discussed earlier to see how affect and prosody is at work in the classroom interactions. If we focus on the soundscape of the classroom – the sounds, voices and their auditory gestures – what might we learn? At first blush, we hear a clear male voice, speaking smoothly, with little stuttering or changes in pitch. Zooming out from this particular performance to the whole-class context of the lesson, we now examine the episode in terms of the multiple sounds that can be heard in the classroom. We focus on what sound is recorded (already a filtering process), so that we might better follow the pulsing forces of the soundscape and its link with other materialities. We are interested more in the combinations of sounds, the peaks and valleys of the pitch levels, and the force of silence, synchrony and dissonance. We agree with Gee (2005) and other discourse analysts who suggest that transcripts should use various symbolic notation to better encode the non-verbal dimensions of discursive practices. This often entails a meta-code of slashes, double slashes, underlines, double brackets, etc., all intended to signify various facets of speech, such as pauses and raised pitch or volume. These techniques often make transcripts harder to read, but perhaps this resistance to the eye is part of the point; perhaps the more diagrammatic transcript might underscore the tension between the different temporalities of speaking and writing, directing our attention to the distinct temporal modulations of speech – melody and rhythm – that are lost when the eye scans the written words at its own pace. Attention to this aspect of speech problematizes our standard transcribing practices in educational research, and perhaps even troubles the very act of transcription. Various 'poetic transcription' strategies have been explored in arts-informed research, offering alternative perspectives on how speech might be studied (see Gee, 1991; Staats, 2008; Staats & Batteen, 2010). Despite these strategies presenting powerful insights and often having more 'face validity' (Lather, 1986), they are rarely treated as reliable forms of representation in mainstream educational research. With an awareness of the impossibility of our task, we proceed nonetheless to attempt to convey the melodies and rhythm without the help of symbolic notion.

We begin at the point where Colin has finished explaining his method using the diagram on the overhead projector. Eighteen seconds into his explanation, the intensity of sounds changes as the quick tapping of the chalk on the blackboard can be heard accompanying Colin's voice. The previously irregular melody of Colin's voice is now overlaid with the louder, quick and constant beat of the chalk scratching through which the teacher paces Colin onward, subtly indicating her presence, her listening. The tapping and talking end at the same time – the teacher-Colin-chalk assemblage stop in synchrony – and are followed by two seconds of silence. The voice of the teacher enters, creating a new intensity that starts with the praising of Colin. The praise is marked both by an increase in volume when she says 'I really like' and a surprising new modifier – the word 'visually', preceded by a short pause and delivered with emphasis – which is surprising in the sense both of being new to this classroom interaction, and also in the sense of its paused, emphatic uttering. Without a skip in beat, the teacher immediately follows the praise by asking how many students understood Colin's method. After a short pause, as she looks around at the raised hands, the teacher nods, saying 'wow, great, okay' and, in the next breath, asks for another method. This is followed by a silence as she looks around the room, deciding which student to call upon. She finally settles on Joe.

The particular rhythm of this transcript is punctuated both by the numbers being counted for each row – 'ten', 'nine', 'nine too', 'eight' – and by the 'one' that is overlapping on each edge. In other words, the rhythm is in alternation of statement and evidence (except for the last two lines, which are both needed to account for the two overlapping 'ones'), which evinces not only a methodical, almost iterative, move through the four sides of the diagram, but an emphasizing of the words that the teacher might use on the blackboard. Further, the initial hedges ('so I, like, um' and 'I just did') give way to a rhythm of gesture. While one might focus on the modality of the sentences in terms of the speaker's level of confidence or certainty at each turn, we hear a voice being taken up by the emergent sound rhythm. That rhythm breaks slightly only with the 'but and then', perhaps intervening now to signal a change in the structure of the argument for the final edge.

It is easy to read this situation in terms of one interval between the teacher and Colin, which ends when the teacher calls on a new student. This reading is supported in part by an interest in the thinking of individual students and in part by a focus on particular mathematical methods. If we focus on the sound-driven blocks of intensity, however, we see a 22-second block of Colin's talking and teacher tapping followed by a new block when the teacher transitions from Colin-talk to Joe-talk. The first block intensifies near the end with the combined tapping-talking assemblage, resolving into a short two-second silence. We see this assemblage of meaning as being essentially composed of human voice and chalk taps, in addition to the diagram, the moving hand and its shadows on the screen, as well as the many other aspects that are not captured in the video. The sound of the chalk announces that the talk and gestures are being inscribed on the blackboard, creating a new material configuration in which the teacher translates and legitimizes the 'method'. The new block begins with an emphatic closure on Colin, where the teacher's voice is actually directed to Colin, and then is followed by a turn – at the same pitch, calm and affirming – to the class, and then poses a follow-up – projected and quick (less than one second) – 'How about a different method?' that emphasizes the sound 'di'. The turn in voice folds the individual talk and action of Colin, along with the teacher's praise, into the common space of the classroom, in which new voices are invited to stand out. The ensuring short silence closes this block, building anticipation for the next event. While watching the video, one can separate the locations of Colin, the teacher and the rest of the students. Each of these easily becomes a source of, as well as an audience for, sound-making. However, a more continuous, relational listening focuses less on each agent. It hears the voice moving between Colin and the diagram being projected, and then meeting and *meshing* with the tapping between the teacher and the blackboard, so that all these different voices merge in sound, despite the geographic distance. We hear the way in which voice travels across the room, first gently settling upon Colin and then sweeping across the room to include the others, as well.

We now skip forward several minutes, during which different students are called upon to explain their methods. We have chosen this particular point in the lesson because it illustrates a very different tonality that points to the different diffractive patterns of sound waves as distinct students become individuated. Of interest also is the way an absent student – Zach, from the class in the previous period – becomes a non-speaking part of the learning assemblage. Once the teacher mentions his name, several voices can be heard talking about him, bringing him into the fold more fully. But before their various comments gain momentum, the teacher breaks in, dispersing this local Zach-talk by speaking and writing on the board at the same time to show Zach's method. Students are now silent as the concert of voice and chalk announces a change of rhythm and a call to attention of both ears and eyes – to all eyes and all ears. She then turns around and says that she will be proceeding backwards and asking the students to explain how Zach's method makes sense in terms of the diagram on the projector. One girl pipes up that she knows another method, but the teacher turns back to the board to finish writing Zach's method – just the sound of the chalk on the board can be heard. With only the chalk to listen to, attention is drawn closely to the marks on the board. When the teacher finishes and turns around, letting a few moments pass before saying aloud the expression 'four times eight plus four', she then asks why the expression makes sense in relation to the picture. Slowly but surely, like a rolling wave, the volume of human voices increases in the classroom, starting here and there like a rumbling and gradually converging into a unified sonic affirmation as students' hands go up at different beats, and the collective sound is punctuated by exclamations. This stretch of time is unique in its polyphonic nature; not only are many voices heard in concert, but they can also be differentiated so that each 'I' that comes to understand breaks out from the group dynamic that has hitherto been punctuated only by single speakers (the teacher, and then each one of the students who was invited to explain their method). At the peak of this growing intensity, after a few laughs are heard, the teacher calls on Kayla to explain why the method makes sense with the picture, thus gathering all those individuating voices into a whole audience for the solo performance. Kayla's voice begins tentatively and is interrupted quickly by the teacher's invitation to go to the board. She continues to speak for a few seconds, before rising and approaching the board. No one speaks as she makes her way to the front and the surrounding silence is sustained as she begins once again to explain, pointer in hand.

Our aim here is not to fetishize speech as the immediate expression of embodied presence, but in fact to recombine speech and thinking in new, material ways and to show how this recombinatorial logic operates in classrooms. Auditory gestures of speech – like rhythm or elision – are often preconsciously absorbed by bodies, as though they were forces of affect rather than discrete percepts that might be isolated and recognized. These forces move across the surfaces of the learning assemblage and are taken in or refracted, and possibly even replicated, without reflection or other rational acknowledgement. The prosodic aspects of speech destabilize grammar and the fixity of lexicons, plugging speech into other materialities that together sustain the event-nature of learning. Prosody is one facet of the affective traffic of the body, a flow of intensity across the human body as it trembles, cavorts and folds into other similarly kinematic bodies. Prosody helps us grasp how gesture operates within speech.

In the next chapter, we focus on the way mathematics has shaped and has been shaped by both the bodily senses and, especially, by changing assumptions about which senses are considered to be more mathematical. This focus is motivated by our general interest in theories of embodiment, as well as our concern that such theories may work to reify the mathematical body in ways that unduly fix the kind of body that can be considered mathematically inventive.

6 The sensory politics of the body mathematical

The writing of this chapter emerged out of our concern with the way in which theories of embodiment in mathematics education almost all assume a typical, or even 'natural', human body. If the body is now going to play such a central role in thinking and learning – whereas before the 'mind', or perhaps even 'intuition', had been the main protagonist in mathematical achievement – what are some consequences for those whose bodies are not typically organised and abled? At a very basic level, we felt that new theories of embodiment had political and aesthetic dimensions, not just epistemological/ontological ones. This chapter focuses on mathematics and the senses. By 'senses', we refer to the sensory organs (eyes, ears, skin, nose and tongue) that we normally associate with our sensing of the external world (hearing, seeing, touching, smelling and tasting). There is quite a long tradition of associating success in mathematics with various senses or with particular sensory organs. By examining this tradition, we can learn more about how conceptions of mathematical (dis) ability are tightly bound to particular images of mathematics and about the relationship between thinking, the body and learning. To this discussion of how various senses have been tied up with mathematics, we bring inclusive materialism to bear in order to rethink the nature of sense altogether. Our goal is to think the body free from the confines of current regimes of perception, and to recognize the human body in all its potentiality, even in our current classrooms, where bodies can be seen as differently abled and differently (organ)ised rather than disabled or distracted. In keeping with our post-humanist approach, we do this by decentring the human organs and abilities – with their fixed forms of sensation, prescribed patterns and implied (dis)abilities – so as to understand perception as distributed across the learning assemblage – occurring in temporary, contingent encounters.

We begin by focusing on the sensory organs themselves, tracing the way that the eyes and ears, in particular, have figured in discussions of mathematical sense. While taste and smell have rarely, if ever, been associated with mathematical sensing, touch has become increasingly pertinent, particularly with the advent of touchscreen technologies.[1] In discussing an example of a young student engaged with a touchscreen device, we show how it is possible to describe this learner's mathematical activity less in terms of the sensory organs she uses (or fails to use) and more in terms of modulating rhythms of sensation. We then explore how notions such as sensation and perception, which have been used by other researchers to describe embodied cognition, have been taken up, following Kant, in terms of the fixed abilities of an individual rather than the underlying hubbub of sensation. With the help of Deleuze's approach to sense and sensation, we propose a rethinking of perception that decentres the organs, while studying the provisional and indeterminate mêlée of sensation.

Following this exploration of the senses and their relation to mathematical intuition, we shift our attention to current conceptions of (dis) ability within educational discourse, charting how conditions such as dyscalculia and autism reveal complex assumptions about the nature of mathematics and of learning. We use the word (dis)ability to underscore the ways in which disability is inscribed within ability, and vice versa, as well as to draw attention to the cultural variability of such delineations. We then draw on (dis)ability studies more broadly to explore political moves that rethink the nature of (dis)ability, including the questioning of the very borders that distinguish the abled from the disabled. We advance the notion of a post-humanist pedagogy that might better attend to the potentiality of the human body.

Mathematics, the senses and intuition

Mathematics has had a complex relationship with the senses over history. This relationship has been shaped in large part by how the senses have been seen as connected to knowledge, to morality and civility, and to spirituality. As Riskin (2002) shows, questions about the senses were very important in

[1] In a similar vein, Jackiw and Sinclair (2009) examine the possibilities that sound might play as a mathematical medium for digital technologies, arguing that it is much more viable than taste or smell, because generators and receptors for these latter two senses are not part of present computer hardware. Sound generators and receptors are, of course, part of present computer hardware, but sound also has a long tradition of association with mathematics, as in the quivering strings of Pythagoras.

scientific and philosophical circles throughout previous centuries, proving to be of particular interest to those who were exploring what it might mean to know and how knowledge might relate to emotions and to values.

Blindness played an important role in debates about the sensory origins of ideas. 'Molyneux's problem', posed in 1688, asked whether a man blind from birth would be able to recognize objects visually were he suddenly to acquire sight. The question fuelled a great debate about the role of the senses, and became a centrepiece in developing and promulgating Locke's empiricist epistemology. Some scholars answered the question negatively, others positively, but a central point of concern was the extent to which knowing with one sense was related to knowing with another. If the blind man had known a cube by touch, would he recognize it by sight?

Berkeley fiercely rejected the claim that the mind could abstract an idea common to all the senses, arguing that there was no reason to believe in a *sensorium commune* – that is, a *common sense* – here used *not* to mean an 'ordinary' and 'unremarkable', or even 'shared', sense across a group of individuals, but rather a unity across one person's different sensory interpretations. He claimed that figures 'perceived by sight are specifically different from the ideas of touch, called by the same name' (Berkeley, 1709, cited in Riskin, 2002, p. 28). Sight and touch were incommensurable for Berkeley, and thus the absence of one sense could not entirely be accommodated by or corrected through another.[2] The debate also featured discussion about the unreliability of the visible world and the relative importance of each of the senses.

With the development of cataract surgery in the mid-eighteenth century, the speculative arguments of Locke, Berkeley, Newton and others became potentially experimental. For the most part, those who had answered Molyneux's problem negatively were right, because patients who were suddenly able to see were bewildered by the visual world, not knowing where to look, not knowing what they were seeing. Furthermore, they believed that what they saw was touching their eyes. In other words, patients could not immediately correlate objects with tactile images: They had no common sense. New questions thus arose: Could patients pass from sight to touch by means of reasoning? Could they 'see' the external world by means of sensation alone? Thus ensued more thought experiments, like those of Condillac, who imagined adding senses to a statue and trying to find out at what point the statue might be said to have sensibility.

[2] Of interest is the work of the multimodal semiotician Jay Lemke and others who take this approach to diverse semiotic modalities in the science classroom.

He concluded that the statue would have sensation without touch, but it would not be able to discover its own boundaries and the existence of the world beyond; without a sense of tactile encounter, there was no determination of a subject.

Mathematics played an important role in these discussions about sensibility. According to Diderot, mathematicians were rationalists who were inwardly focused. This made them ultimately insensible to the external world; they were like blind men because of their solipsistic, Cartesian casts of mind. Indeed, it might even be considered advantageous for a mathematician to be blind so as to avoid deception by ocular tricks.[3] Diderot derived these conclusions partly from reading the memoir of, and his subsequent interviews with, Nicholas Saunderson, a blind professor of mathematics at Cambridge. But these conclusions also stemmed from his belief that the development of one's moral faculty depends on the extent to which one was affected by external things. Because the senses were presumed to be the primary means of being affected emotionally, rationalists were less moral. And mathematicians were considered rationalists by nature and by practice, in part because of their insensibility to affect, which meant they were less than perfect human subjects. Those involved in these experimental discussions shaped an image of the perfect human subject as combining physical receptiveness with emotional and moral openness. Indeed, as Riskin (2002) argues, Locke's original philosophical position that ideas are composed from sensations gave rise to a new 'presumption that knowledge resides in a combination of sensation and emotion' (p. 20). In other words, knowing meant not only sensing, but also feeling; one must become sensitive to the existence of external things; one must grasp that these things are indeed external to oneself. This was an epistemological position that bound together the moral and the physical.

But before this eighteenth-century interest in the sensory organs, we find a much older connection between mathematics and both affective and physical perceptiveness. In the thirteenth century, Henry of Ghent spoke of 'the melancholy disposition of the mathematical mind', wherein melancholy arose from an obsession with the structures and forms embedded in actual matter. According to Henry of Ghent's view of mathematical activity, mathematicians inevitably think of things that 'must have extension or, as the geometrical point, occupy a position in space'; mathematicians make the worst metaphysicians, 'for they cannot raise their minds above

[3] Anamorphosis – drawings made using devices that distort perspective – plays a part in this story of eighteenth-century suspicion of the visual.

Figure 6.1. Melencolia I, 1514 (engraving), Albrecht Dürer (1471–1528)/Private Collection (The Bridgeman Art Library).

the spatial notions on which mathematics is based' (in Klibansky, Panofsky, & Saxl 1964, p. 338). The association between melancholy and mathematics continued into the sixteenth century, as evidenced by Albrecht Dürer's famous engraving *Melencolia I* (see Figure 6.1), which is widely regarded as an expression of this disposition, with its many mathematical elements, including a magic square, various three-dimensional solids, and a compass and straightedge. However, the later 'glorification of melancholy' in nineteenth-century Romanticism involved a very different notion of melancholy than that of Henry of Ghent. This *romantic* view of melancholy posits it as a condition of the soul, which, according to Yates (1979), involves the *starving* of bodily senses, which then occasions the 'intense inner vision' that is often associated with acts of creativity or experiences of intuition. If Henry of Ghent's melancholy stemmed from an investment in the sensory world, modern melancholy required withdrawing from it (see Sinclair & Pimm, 2009).

While Henry of Ghent, Diderot and others have had varying and perhaps idiosyncratic views on the relationship between mathematics and the senses, the influence of Kant on contemporary conceptions of this relationship is both indisputable and long-lasting. Kant offered a way of talking about mathematical knowledge, and its relation to the senses, in terms of intuition. He distinguished between intuitions and concepts, where intuitions were 'singular mental representations of particular objects' and where concepts were 'general notions used to classify and sort objects' (Bostock, 2009, p. 46). There are two kinds of Kantian intuition – one that furnishes immediate knowledge through the senses, and a second that yields *a priori* knowledge of necessary truths. According to Kant, it is this pure intuition that supplies mathematical truths and delivers the *forms* of possible empirical intuitions. Indeed, in the synthetic *a priori*,[4] Kant carved out a new philosophy of mathematics that centred the human mind and its ability or faculty to evaluate mathematical truth.

The Kantian legacy is the belief that mathematics is the mind's tendency or faculty for apprehending the *forms* of perception. We see in this legacy the birth of a persistent belief that mathematical intuition springs from a mental faculty or innate ability of the human mind. Of course, Kant was not the first to propose an innate mental faculty for doing

[4] Kant claimed that mathematical truths are uniquely *synthetic a priori*. Mathematical truths are *a priori* in being both necessary and universal, and they are synthetic – rather than analytic – because they entail material implication. For example, an analytic statement like 'all bodies are extended' conveys nothing more than that which is contained in the Cartesian concept of 'body', whereas the synthetic statement 'all bodies are heavy' asserts a truth that reaches outside the given concept. The statement 'The straight line between two points is the shortest' is a synthetic proposition, because, according to Kant, 'my concept of straight contains nothing of quantity but only of quality' (Kant, B16, quoted in Bostock, 2009, p. 49). Kant also considers this statement as *a priori*, because he believes its truth cannot depend on any sensory experience.

mathematics – Descartes believed that mathematics was unlike all other knowledge in being absolutely certain, and thus the truths of mathematics could not be truths about the physical world, nor could they depend on perceptual experience – but Kant's project was the most compelling in how it reckoned with perception, or the synthetic aspect of mathematical truth.[5] Like Kant, the intuitionist philosopher of mathematics L. E. J. Brouwer believed that mathematics was synthetic *a priori* and that mathematical truths could be known neither through the senses, nor through mere analysis of the concepts involved – one needed intuition to go beyond the concepts to determine their truth-value. For Brouwer, as for many who follow in this Kantian tradition, it is not the senses but rather the human faculty of intuition – revamped post-Kant to address the existence of non-Euclidean geometries – that determines the truth of mathematics.[6] In Brouwer's words, the first act of intuitionism:

> completely separates mathematics from mathematical language, in particular from the phenomena of language which are described by theoretical logic, and recognizes that intuitionist mathematics is an essentially languageless activity of the mind having its origin in the perception of a move of time, i.e., of the falling apart of a life moment into two distinct things, one of which gives way to the other, but is retained in memory. If the two-ity thus born is divested of all quality, there remains the empty form of the common substratum of all two-ities. It is this common substratum, this empty form, which is the basic intuition of mathematics. (Brouwer, 1952, pp. 141–42, quoted in Brown, 2008, p. 121)

Intuitive is one of those slippery terms that are often used to describe a quality of students who seem to be gifted with mathematical talent. The history of mathematical intuition is strongly linked to a belief in the innate cognitive ability of the human subject and a corresponding disbelief in the senses as a source of mathematical knowledge. We have highlighted how

[5] According to Descartes, a mathematical truth is based on innate, clear and distinct ideas. Historians of philosophy usually contrast Descartes (a Rationalist) with Locke and Hume (Empiricists). However, Bostock (2009) claims that Locke, Hume and Descartes, despite their differences, share the same beliefs about the ontological status of mathematical objects (that they are ideas, or mental entities) and suggests that these philosophers differ only in how they think we acquire these mathematical ideas.

[6] Such a perspective is evident also in Hadamard's (1945) work on the psychology of mathematical invention, which features intuition quite centrally. He wrote, however, that intuition needed to be legitimized by rigour: '[T]he object of mathematical rigor is to sanction and legitimize the conquests of intuition, and there was never any other object for it' (in Pólya, 1981, p. 127), which gives intuition a kind of wild-child status, emerging as it does – according to him – from the depths of preconsciousness.

the concept of intuition is bound to a Kantian heritage in the philosophy of mathematics and how it emerged historically alongside an eighteenth-century belief that the (dis)abled mathematician's body – through blindness or melancholy – afforded the mathematician stronger access to the unsullied insights of this intuition.

This brief history of the role of the senses in conceptions of mathematical (dis)ability reveals the centrality of the eye as a mathematical organ, as well as the almost mythical role that intuition plays in somehow bridging the senses and reason (for Kant) and/or in accounting for mathematical invention and creativity (for Brouwer, Hadamard, Poincaré, etc.). However, intuition has no organ of its own, but like the sense organs, it is often assumed to vary along a range from full capacity to total absence. Thus the concept of intuition plays a powerful role in the politics of mathematical (dis)ability, and is often discussed in the context of poor pedagogical approaches that fail to nurture student intuition (see Burton, 1999). As we move to the context of school mathematics, the organs central to current concern might be the eyes, the ears and the fingers, but in the next section, we argue that these senses must be considered provisional. By focusing on how the senses are configured in the moment of an encounter when new bodily assemblages are formed, rather than being given in advance with a prescribed sensibility, we hope to de-naturalize intuition and trouble the taken-for-granted emphasis on certain senses. We show how such an approach sheds light on a video excerpt of a five-year-old girl and a teacher working with an iPad.[7]

Mathematics education and the sensory organs

In their recent chapter in the *International Handbook of Mathematics Education*, Healy and Powell (2013) draw on Vygotsky's stance on organs (the eye, the ear, the skin) as tools that can be considered in much the same way that material instruments, language or other semiotic resources can. While hearing children use their ears first to learn about numbers (especially to chant the numbers up to 10, for example, but also to 'overhear' the constant buzz of everyday quantifying), children who cannot hear can use their eyes as tools to learn the same concepts. From our inclusive materialist approach, this metaphor of tool seems to re-establish an inside/outside distinction that is fixed – the ears, eyes and skin are part of the inside, and

[7] There is not enough space here to discuss how video data itself selects and limits the kinds of ways we might study the senses and sensation in the classroom, although such a discussion is extremely important.

what is heard, seen or touched becomes de facto outside. Tools are treated as conduits or mediators between outside and inside, and concepts are constructed through mediation but are ultimately assumed to be universal. Instead we want to think of organs as part of an assemblage that emerges at a given time but does not precede the phenomena, nor submit entirely to human control. In displacing the sense organs in this way, our aim is to show how sensation operates below and sometimes transversally to that which is recognized as perception. This may seem 'counter-intuitive' to some readers, but we believe that such an approach allows for a new politics of dis/ability. We illustrate our approach through an analysis of a brief video excerpt; in the analysis, we also introduce some ways of speaking about learning and sensing that we will return to in later sections of this chapter.

We have chosen to exemplify this approach to sensation using an encounter between a child, an adult, a touchscreen device and a concept of number. Touch technologies are currently lauded for the way they are meeting the needs of learners with special needs; various mathematical and educational applications have been developed for these devices. In discussing this video, our aim is to offer an alternative interpretation of processes of learning number without foreclosing on the futures of the body. In other words, instead of focusing on what specific organs can or cannot do – as if they were mediating tools – we suspend judgement on those organs, loosening the boundaries on their predetermined functions and spatio-temporal qualities. We examine instead the way that waves of sensations of varying intensity move across the learning assemblage and the way that the encounter between teacher and student and screen entails complex rhythms of participation.

We have also selected this example because it underscores the way that new media (in this case, the iPad) entail new sensory modalities and, in our estimation, new configurations of sensory organs and mathematical concepts. The sense of touch is emerging as quite central in the literature on embodiment in mathematics education. As Roth (2011) writes, even for Aristotle, touch was the "sense that distinguishes humans from animals" (p. 51), accounting for the former's intelligence:

> For in the other senses man is outstripped by many of the animals, but in point of touch his accuracy exceeds that of the others by a long way. And it is also for this reason that he is the most intelligent of the animals, an indication of which is the fact that even within the race of men it is in accordance with this sense-organ that individuals are well or badly endowed by nature, and in accordance with no other. (Aristotle, 1986, p. 180)

As discussed in Chapter 1, Roth also focuses on the sense of touch, which he takes as central to learning. He is interested in the way touch directly affects and is affected by the surfaces it explores, like the child holding the cube – he touches and is touched by that cube. The sensation of touch necessarily involves movement; as Merleau-Ponty (1945) writes, 'it is the manner in which the surface uses the time of our tactile exploration to modulate the movement of our hand' (p. 364). Touch also becomes an important way of enabling blind students to link up with a mathematical solid, feeling the slope of its sides, the hardness of its edges and the pointiness of its vertices (Healy & Fernandes, 2011). But touch seems somewhat restricted in relation to mathematics, given that most mathematical objects are not usually thought of as being accessible by touch.[8] The growing use of touchscreen technologies, however, as well as the increased focus on gestures, is changing this situation. Not only is it now possible to touch various mathematical objects on the screen, such as functions, numbers and geometric objects, but the gesturing hand (especially the finger), as it traces a curve in the air, say, might be said to produce virtual touches. These technologies may well herald a twenty-first century of touch-centrism that displaces the current era of ocularity.

The mathematics classroom, especially at the elementary school level, has long employed a variety of material resources to support student learning. These 'manipulatives' are often used to provide concrete forms of interaction, so that children can actually touch three-dimensional shapes or work physically and visually with blocks, abaci or geoboards. The advent of computer screens in the early 1980s helped promote new visualizing habits in the higher grades, as well, inviting learners to operate through the use of a mouse and keyboard. More recently, the introduction of touchscreen devices (such as interactive whiteboards, iPads, tablets and other similar hardware) has enabled tactile interaction with the screen, so that the fingers and hands, through touch and gesture, directly create and transform mathematical objects. The direct contact can break down the representational epistemology that is often taken for granted in mathematics education, offering the sense that one is actually touching the mathematics and not some screen simulation. We are led to ask: How does the sense of

[8] Actually, Ivins (1945/1964) has argued for the tactile-muscular nature of Greek geometry, whose two principle assumptions – congruence and parallelism – both involve 'intuitions' about things at hand, be they the chalk line; the straightedge and the superimposing of one triangle on another; or the tracing of fingers along opposite sides of a rectangle. Indeed, the idea that parallel lines never meet is much more tactile-muscular (close your eyes and move your fingers along opposite sides of your screen or your book) than it is visual, given that parallel lines always meet at the horizon in our perspectival world.

touch change when actions of tapping, sweeping and finger-releasing begin to dominate our encounters in what many are calling *screen culture*? And how do mathematical concepts change when they partake of this kind of activity? In the following analysis of a short video excerpt, we pay attention to the way that the learning assemblage entails all sorts of rhythmic folds of sense and feeling; we show how the concept of number that emerges is part of a block of sensations. Thus, our aim is to begin to grasp the intensities and forces that lie within sensation and that draw learners into becomings (of human and non-human bodies). The numbers and figures of mathematics – which are typically considered non-human – partake of this rhythmic sensation. We want to explore what Smith (2003) means, in the context of mathematics, when he says that '[b]eneath concepts, one always finds rhythmic blocks of complexes of space-time' (p. xix). In other words, beneath concepts and geometric figures there is a molecular pre-individual block of sensation.[9] Or as Deleuze would say, 'it is rhythm itself that would become the Figure, that would constitute the Figure' (Deleuze, 2003, p. 60).

The example we discuss here involves a teacher and kindergarten student (aged five) working together with an iPad application called *TouchCounts* that was designed to support the development of number sense, with a special focus on the tactile interaction, but also making use of the visual and speech-generating capacities of the device. We focus on just one part of *TouchCounts*, in which a learner taps her fingers on the screen to summon numbered objects – or objects representing numbers. The first tap produces a new yellow circle on which the numeral '1' appears. Subsequent taps produce sequentially higher numbered objects. As each tap summons a new numbered object, *TouchCounts* audibly speaks the name of its number ('one', 'two'). As long as the learner's finger remains on the glass, it holds the numbered object, but as soon as she 'lets go' (by lifting her finger), the number object falls 'off' the bottom of the screen, captured by some virtual gravity. If the learner releases her numbered object above the shelf, or 'flicks' it above the shelf on release, it falls only to the shelf, and comes to rest there, visibly and permanently on screen, rather than vanishing out of sight 'below'. (Thus, Figure 6.2b describes a situation in which there have been four taps below the shelf – these numbered objects were falling – and then a '5' was placed above the shelf.) Each time a finger is placed on the

[9] It is interesting to contrast the notion of rhythm with that of pattern, the latter being central to our discussion of pattern and mathematical intuition in Chapter 7. Pattern is assumed to be visually perceived (in figures or perhaps alphanumeric inscriptions) and masking some invisible logical relationships, whereas rhythm insists on and is inextricably entangled in time and material vibration.

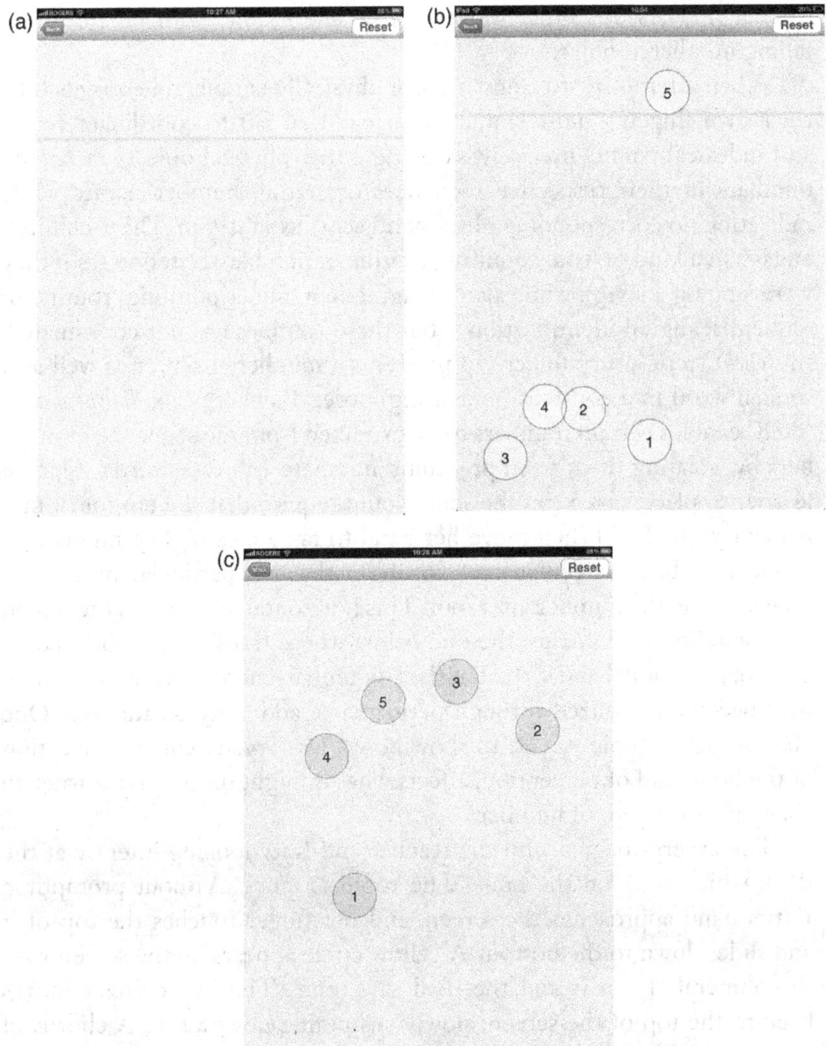

Figure 6.2. (a) Counting world; (b) with a number on the shelf; (c) without shelf.

screen, a new numbered object is created, so one cannot 'catch' or reposition an existing numbered object by retapping it. The block of sensation – sound, touch, sight, command – emerges through the movement of the learner as she taps and sweeps her finger and leans in and out, as well as by the movement of the objects on the screen as they appear and then slowly descend to the bottom of the screen. As we will see, the movement of the teacher mirrors this other movement, as she follows with her head the falling numbered objects.

When attempting to count a set of physically enumerable objects (like their own fingers), most kindergarten children fail to coordinate verbal and indexical counts precisely, skipping either physical objects or named numbers in their respective sequences or stating number names while indicating no corresponding physical object (Fuson, 1988). These children engage in a kind of 'rote counting' routine of number recitation (as if they were singing a song), while also engaging in a 'finger pointing' routine of sequential digital identification – but these routines are not coordinated. In *TouchCounts*, every finger tap produces a numbered object as well as a spoken word in a constant (increasing) order. Furthermore, *TouchCounts's* 'shelf' enables certain numbers to be extracted from the sequence of numbers by isolating them from preceding numbers. In other words, asking a learner to place 'just 5' on the shelf would require that she tap four times below the shelf and then move her hand to tap above it. The number '5' would thus become reified into an object that has particular properties, such as being the number after four. This horizontal line divides the screen into two distinct domains, the one below where transient numbers come and go, provisional and virtual in their potential, and the one above where they become actualized through permanence and fixity on the line. One goal of this example will be to show how a less organ-centric conception of the body, and of perception, affects what it might mean for a learner to change her concept of number.

The excerpt begins with the teacher and Katy looking intently at the iPad, which rests on the table. The room is quiet. Without prompting, Katy's hand approaches the screen, and her finger touches the top of it and slides down to the bottom. A yellow circle appears on the screen with the numeral '1' on it and the iPad says 'one'. The index finger moves back to the top of the screen, slowly swimming downwards. A chorus of 'two' comes both from her mouth and the iPad. This happens repeatedly, although sometimes only the iPad can be heard announcing the new number while Katy's lips move in synchrony (Figure 6.3a). The appearance of '10' on the tenth yellow circle attracts attention, perhaps because of its double digits, and Katy bends over to look closely. Now only the iPad counts the numbers (Figure 6.3b). After 'seventeen', several fingers fall on the screen at once, and then 'twenty-one' is heard. This produces a pause in the action, and Katy's lips spread into a smile. All but the index finger are tucked away, as the rhythmic tapping continues along with the chorus of named numbers. At 'twenty-seven' Katy looks up, no longer watching the

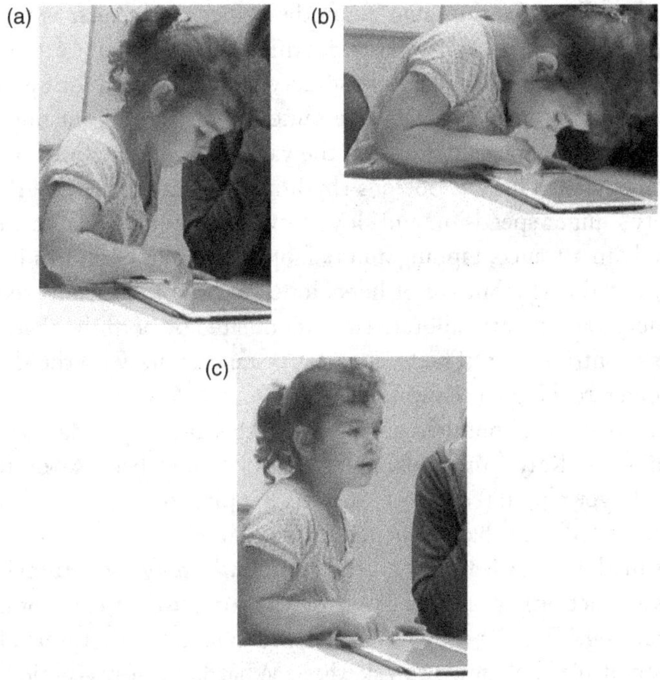

Figure 6.3. (a) Katy's swimming index finger; (b) focus on 10; (c) not watching.

screen (see Figure 6.3c), and she continues swiping and saying numbers. The ears and finger have taken over and are only interrupted when the finger accidentally hits the Reset button.

It is tempting to analyse Katy's first interactions with *TouchCounts* in terms of the distinct sensory forms of feedback that she received, with touch enabling her to produce objects, the numbered yellow circles allowing her to see those objects and the voice of the iPad producing sounds for those objects. In this reading, Katy would have been making sense of each sensory dimension – turning sensation into perception, à la Kant – and then coordinating or synthesizing these discrete parts into a concept of number. However, it is clear that the encounter among fingers, ears, mouth, eyes, as well as iPad, numbers and the teacher entails a far more entangled process where non-sense and sense are bound up together in productive ways. If we take the view that the sense organs are not fixed tools through which sensations turn into perceptions, and we study instead to the rhythm that operates beneath concept-producing judgements, our attention is drawn to micro-level relational responses. Of course, that which constitutes the micro in our study will not be the same in another, since terms such as micro and macro depend on one's apparatus. In this case, we focus on the various tempos and rhythms of the haptic-sonic-visual encounter, for instance the repeated tapping, whether it be the same finger or different fingers, and the tiny delays and accelerations in the various recombinations of speech, sight and touch. There is obvious rhythm in this act of repeated tapping, and Katy's finger speeds up and slows down as she proceeds, at one point seemingly in a trance, tapping and talking as the numbers get higher and higher. But this rhythm is not hers alone, because the various sensations that emerge are in part calibrated and modulated by both the iPad and the teacher's contributions. The teacher, for instance, joins with the rhythm of the tapping, rocking back and forth.

The concept of number emerging in this brief episode has already evolved from Katy's previous experiences, if only because of the new actions (finger tapping) and intra-actions (tapping/seeing, tapping/hearing and tapping/talking) that are being occasioned. However, inasmuch as the goal of mathematics learning is to continuously reify one's thinking (and talking and acting), we might like to see a more distinct concept of number emerging. This, for us, will require a change in the rhythmic blocks of sensation at play. We want to ask, then: What kind of rhythmic block of sensation underlies a particular kind of reification? If reification involves a turning of a process into an object, what might we expect in terms of the new configuration of the finger-screen-voice-numbers assemblage? How might a post-humanist approach describe not just a human/machine interaction, but also the production of a new concept in that interaction? We tentatively answer this question by examining more closely how the finger-screen-voice-number assemblage responds to the teacher's refrain, 'I want to see 5 but not the other numbers.'

Katy's index finger taps above the shelf and the teacher says, teasingly, 'I don't want to see one.' Katy's lips spread briefly into a smile then taps again above the shelf, and the teacher tells her she does not want to see '2', either. The smile returns. The finger taps to make numbered objects '3', '4' and '5' above the shelf, and then presses Reset. Laughter. This laughter marks a new rhythm associated with the Reset button, a button that performs like a big bass drum, clearing the screen at various time intervals, and allowing for the tempo of engagement to recalibrate. Katy's head bends over the iPad, her index finger taps above the shelf, and then taps three times below it. The fifth tap is above the shelf. She sits back and smiles. The teacher says, 'Now I can see 1 and 5, but I just want to see 5.' Katy's index finger presses on the numbered object '1', trying to move it off the shelf, but instead it produces the numbered objects '6', '7' and '8'. Reset. It is important to note here that her finger – as the main organ of touch in this encounter – has just been reconfigured in terms of its capacities. No longer the organ that can only move or drag the yellow circle, it now becomes a finger that resets. The functionality of the Reset button, and – as mentioned previously – its power to recalibrate the tempo and rhythm of this encounter, becomes part of the finger assemblage, thereby redefining what is currently entailed in the sense of touch. One also should emphasize that the sudden and complete erasure of the screen by the Reset button, which triggers laughter from Katy, changes the potentiality of touch quite radically. The power to erase all one's traces, and to erase them all at once, changes Katy's investment in the repetitive act of touching and counting. After pressing the Reset button, she uses her index finger to put numbered objects '1', '2', '3' and '4' below the shelf, while her lips mouth the associated number words. She pauses, and then taps above the shelf. The teacher asks whether she can put '5 and 10 up there'. Katy's finger moves to the Reset button then taps above the shelf, and then rushes to Reset again. Four taps below the shelf are made quickly, and then a pause. Katy's eyes look to those of the teacher, then down again as her finger taps below the shelf. Her eyes follow the numbered object '5' as it falls down. She says, 'Oh!', and the teacher says, Oh no, you lost it.' The finger moves to Reset, taps quickly four times below the shelf, lips moving in synchronization along with the iPad, and then one tap above the shelf.

We stop describing the excerpt here because it is sufficient to show the emergence of Katy and the concept of five: becomes individuated so as to highlight its relation to other numbers (like four). More precisely, the finger-screen-voice-five assemblage changes in such a way that five becomes four finger taps (as well as spoken and mouthed words), followed by a fifth tap that occurs temporally and spatially distinct. The concept of number is alive and changing as this learning assemblage continues to grow. In particular, the cardinal and ordinal aspects of the changing concept of number are linked to the motion and gravity-like force that operate on the screen. Again, it is not enough to say that Katy learns by getting visual or audible feedback from the iPad enough times to *see* or *hear* that five is indeed the number that comes after four. Instead, what is clear from the video is that Katy is taking part in a rhythmic engagement that involves a particular assemblage of gestures, sounds, touches and moving yellow circles. This assemblage is provisional and responsive to the materiality of the event; the potentiality of her sense organs has been altered by this encounter. The haptic-sonic-visual fusion of sensory data emerges in this encounter in unique ways – ways that not only reflect the impact of new technologies on learning, but also reflect how sensation is not contained by the sense organs; rather, it is flowing and unstructured, always open to alternative recombinations.

Rethinking perception and sensation

Kant's theory of perception involves a *synthesis* of dispersed sensation. This cognitive synthesis entails first apprehending and then contracting the sensations into a unified perception of some particular thing. (Smith, 2003). In other words, an object is perceived when it has been assigned to the synthesized parts of a spatio-temporally apprehended multiplicity. This begs the question as to what actually constitutes this original multiplicity. What is this 'sensation' at the outset of the process? How is a sensation recognized to begin with, before being synthesized as a whole? These are the unstated questions of the *Critique of Reason* that get taken up in the later *Critique of Judgement*. Kant explains how pre-perception 'parts' are identified as parts by the imagination – not the understanding – through the use of a sensible or qualitative unit of measure. According to Smith's reading, Kant proposes that our bodies know intuitively how to select the unit of measure. This idea gets taken up and elaborated in later phenomenological treatments of the body. For instance, Smith (2003) discusses how the phenomenologist Erwin Straus (1935/1963), in *The primary world of the senses*, similarly distinguishes between sensation and perception, where perception is a secondary rational organisation of the raw data of sensation. According to Straus, visual sensations are sensed without shape or spatial configuration, whereas visual perceptions are sensed through agential bodily production and synthesis. Straus treats sensations and perceptions as coextensive, operating alongside each other, the one 'intentional' or spontaneous, and the other coordinated in relation to external spatial constraints.

Deleuze (1984) develops this distinction between sensation and perception in a slightly different way, drawing attention to the fact that Kant's pre-synthesis act of evaluation, whereby a sensation is identified, seems to be operating even at the very minute level of apparently immediate sensory encounters. Deleuze suggests that such an evaluative act at the micro-level is relational and highly responsive, and thus the unit of measure is itself in constant variation and infinitely divisible (Smith, 2003). This variation and recalibration is, in Deleuze's terms, a grasping of a rhythm that operates beneath the concepts entailed in judgement. The constant recalibration is less a human synthesis of discrete sensations and more the synchronizing of rhythmic intensities across a system. But synchronizing is radically different from synthesizing. We propose that perception is not the synthesis of sensation – where synthesis is taken to be a rational judgement – but rather a polyphonic process of modulation, a process by which new folds and inflections emerge in unstable material configurations.

The continuous variability of the unit of measure indicates how sensations are more like folds than individuations. For Deleuze, sensation is vibration, and thus rhythm is the foundation of perception. In what is a far more Leibnizian than Kantian approach to perception, Deleuze suggests that perception is not individualized, but rather comprises the rising amplitude of a wave or rhythm, with frequencies momentarily in phase and resonant. This rhythm is meted out in the *relation* between resonant forces – a rippling effect across proximal molecules, a change in pressure or intensity. Both sensation and perception emerge through this molecular bodily calibration, not simply as a human faculty à la Kant, but rather as an aspect of the physical world. Thus, both human and non-human perceptual synthesis is an aesthetic comprehension that is founded on rhythm (synchronizing melody). In Chapter 5, we explored how language use in mathematics classrooms can be studied for its rhythmic qualities, but our aim here is to pursue this post-humanist approach so as to study human perception more generally.

There are obvious consequences for the body when we pursue this approach to perception. Like Barad's discussion of the cat thought experiment (described in Chapter 2), 'life' is stretched in all sorts of new directions through a plurality of invisible forces that are reshaping the individual. This is no longer a body with determinate organs; the body is now defined by the provisional and temporary presence of indeterminate organs:

> A wave with a variable amplitude flows through a body without organs; it traces zones and levels on this body according to the variations of its amplitude. When the wave encounters external forces at a particular level, a sensation appears. An organ will be determined by this encounter, but it is a provisional organ that endures only as long as the passage of the wave and the action of the force, and which will be displaced in order to be posted elsewhere. (Deleuze, 2003, p. 41)

Sensation is an elastic and modulating wave of intensity flowing through space, and the body is that which sustains the sensation and the provisional congealing of organs.[10] The body is thus constituted through thresholds, or levels of resonance, and sensation is dislocated and only provisionally situated in a perception.

This is a body with potentially different perceptual capabilities than what are currently considered normal. Hearing, touching, tasting, seeing and any other modality are temporary – if persistent – kinds of perception, but our future may entail entirely different calibrations of sensation, as we hoped to make evident in our description of Katy and the iPad. Thus, the question 'what can a body do?' is a pressing one, because the answer is not given in the current bodily arrangement. According to Deleuze, the body is always in a process of becoming, and potentially becoming perceptually enabled differently. The future of a body is unscripted, the sense organs unfinished and open to reconfiguration and dispersal. It may seem implausible to some readers, given the long evolutionary process by which the ear, for instance, has come to be organised in its current form, but given the proliferation of new devices for augmenting our sensory engagements with the world – see for instance, Pranav Mistry on sixth-sense technology – the future may indeed offer new kinds of sensory organ arrangements.[11] This post-humanist approach to sensation allows the body to break free from the confines of current perceptual organisation and demands that we recognize the human body in all its potentiality, even in our current classrooms, where bodies can be seen as differently abled and differently organised rather than (dis)abled or distracted.

Discussions of mathematics and learning (dis)abilities are difficult in part because one is always working against the Kantian tradition – and our inheritance of this tradition in the still-dominant paradigms of cognitive psychology – in which ability is ascribed to some sort of internal intuition or mental faculty. Such an approach continues to look suspiciously at

[10] An artist like Francis Bacon captures the proliferation and distortion of the senses in the human body. In doing so, he moves away from the figural nature of classical portraiture and away from a purely optical code for painting. Deleuze likens Bacon to Cézanne in his attempt to paint the sensation itself rather than the figure or the narrative situation. For both of these artists, sensation is not the free play of disembodied light, but literally the force that fuels the becoming of a body. 'What is painted on the canvas is the body, not insofar as it is represented as an object, but insofar as it is experienced as sustaining this sensation …' (Deleuze, 2003, p. 32).

[11] See http://www.pranavmistry.com/projects/sixthsense/.

material assemblages of hand, eye, paper and concept that are the occasions of learning and inventing mathematics. We have shown, though, how the meaning and value of the senses have changed over time, and how this sensory politics has influenced the history of mathematics. Such politics are at work in any claim that certain kinds of senses might be more ideal, more sensitive, more subjective or more objective.

Such a sensory politics is also at work within the discipline of mathematics itself. This is obvious from a broad historical perspective in that the oral tradition of ancient Greek mathematics differs quite radically from the written tradition of contemporary mathematics (as we discuss in Chapter 7 in relation to Netz's work). But it can also be seen more recently in the early to mid-twentieth-century priorities of the Bourbaki group, in which mathematics was first and foremost a matter of transcendental structure and the representation of that structure by the alphanumeric, whether spoken aloud (mathematicians would read drafts of books aloud to each other in small groups) or written. Thus, the role of vision narrows to the alphanumeric, and the image is outlawed (see Pimm, 2006).

However, if Rotman (2008) is right in suggesting that the future of mathematics will see a reduction of an alphanumeric hegemony and, in particular, a move towards visual and dynamic mathematical expression, there will be a new kind of sensory politics at play. We might see the entanglement of new media and mathematics as a move towards accessibility, but we will also need to ask what the consequences of the new politics will be: For example, might the blind (mathematician or learner) be newly (dis) advantaged? What will be the upcoming demands on the moving body? Answers to these questions will depend on the very ways in which we construe (dis)ability.

In the next section, we relate this approach to questions of mathematics learning and (dis)ability. Our aim is to elaborate on the challenge we have brought to organ-centric views of the body, and explore the implications of this approach in the context of school mathematics.

The sensory politics of (dis)ability in mathematics

The final three decades of the twentieth century saw considerable debate about the existence, definition and identification of (dis)abilities (Brantlinger, 2001, 2004; Woodword, 2004). Borgioli (2008) distinguishes between a modernist (medical) approach to (dis)ability, which assigned its source within the individual, and a postmodernist approach that views 'disability as a social construction based on incorrect, immoral assumptions regarding difference' (Hallahan and Mock, 2006, p. 27). The former approach continues to dominate the field, as is evidenced in the U.S. *Individuals with Disabilities Act* (IDEA, 2004), which explicitly defines disability as a disorder *within* the individual. The latter approach has gained support from neurophysiological research that underscores the plasticity of the brain. In line with the second approach, Borgioli (2008) argues that learning disabilities in mathematics – and this would include disabilities such as dyscalculia – are constructed using narrow definitions of what counts as acceptable mathematics and what counts as evidence of proficiency in mathematics. In practice, many students labelled LD in mathematics receive highly directed step-by-step instruction of basic skills and procedures (Baxter, Woodward, Voorhies &Wong, 2002; Fuchs, Fuchs, Hamlett & Appleton, 2002). This focus persists whether these students are pulled out for special education interventions or are offered differentiated instruction within regular classrooms (Hehir, 2005; Woodward & Montague, 2002). Typical kinds of instructional strategies offered for LD students – such as grouping similar problems together for the sake of easy recognition or coding operations in colour for easy association – reveal assumptions about mathematics that are at work in these pedagogic strategies. Such strategies might help students master procedures or recall an algorithm, but they fail miserably at cultivating any kind of adventure and passion.

The 'dyscalculia' label provides an interesting case study of the desire to ascribe learning difficulties to individuals, rather than to any particular socio-political-technological assemblage. Indeed, as Munn and Reason (2007) point out, the notion of dyscalculia can be used to 'explain', in objective neuroscientific terms, the low performances of certain learners who have been identified as experiencing difficulty participating in certain practices of school mathematics. Healy and Powell (2013) note that these practices are almost always arithmetic-based and, moreover, assume a paper-and-pencil technology for number operations that values procedures (such as adding two-digit numbers by placing them vertically on the paper and 'carrying' numbers from one column to the next) over the exclusion of other more holistic or contextual strategies. Because the term dyscalculia has migrated into mathematics education discourse, one might assume that the condition it describes actually exists, even in the anaemic form of 'faulty' procedural reasoning about number operations. But Gifford (2006) questions whether dyscalculia describes a specific cognitive deficit, in part because no consensus has been reached about its defining characteristics, aside from poor recall of number facts. While such a consensus may well emerge, as might neurological differences between individuals in their processing of number, any exclusively neuropsychological approach is highly problematic.

The ongoing struggle that both deaf and blind students have been facing in the mathematics classroom over (at least) the past century underscores the need for new approaches to thinking about these sensory (dis) abilities. On the one hand, the implicit way in which the materiality of (school) mathematics is ignored – so that mathematical concepts are treated as independent of the ways in which they are represented, communicated, explained or understood – functions to devalue the role of the senses in learning. On the other hand, the actual sensory ways in which mathematics is represented, communicated and explained tacitly privilege certain sensory capacities.

Decades ago the research asserted that these students' respective sensory losses were the cause of failure in mathematics, but new researchers have pointed to the particular sensory politics at play in mathematics classrooms that create barriers to learning. While there are many kinds of such barriers (for example, deaf and blind learners are less likely to be taught by mathematically trained teachers), there is a growing awareness of the way in which certain sensory modalities are privileged, often only implicitly, in school mathematics. There are, of course, obvious ways in which assumptions about sight and hearing are embedded in classroom practice, such as the privileged role of the blackboard and the extensive use of verbal instructions and explanations. Even efforts to provide alternative forms of access remain problematic, in that they impose additional mathematical or, more often, memory requirements on the learner than the original task did. In seeking to understand how deaf and blind learners develop mathematical expertise, contemporary researchers have made evident two important issues: (1) the loss of one sense may change the way other senses are used, which may lead to certain opportunities that often go untapped[12]; and (2) mathematics itself changes under different sensory organisations.

In terms of the first issue, Marschark, Spencer, Adams and Sapere (2011) stress the fact that deaf and hard-of-hearing (DHH) learners are *not* 'simply hearing children who cannot hear' (p. 4). Rather, they have specific

[12] It is important to note that such a claim is not equivalent to an 'accommodation' approach, whereby one sense takes up the absence of the other. Rather, we emphasize in this claim that the sense organs are plastic, or open to mutation, even within the lifetime of the individual.

cognitive and language abilities that can make some kinds of mathematical ideas harder to learn and others easier. For example, in her book *Teaching Mathematics to Deaf Children*, Nunes (2004) carefully shows how the seriality of speech (and hearing) can make it more challenging for deaf learners to learn how to count if they are not provided with alternative modes of instruction and forms of experience that provide visual spatial presentations. But while deaf children often underperform on counting tasks, their more spatial way of thinking about number allows them to perform better than hearing learners on tasks such as counting backwards and 'What number comes after this one?' But these are not the tasks that are used to establish norms around number sense. It is in this sense that mathematics itself – or, more precisely, school mathematics – is formatted by certain sensory assumptions.

There is simply no objective basis on which, as educators, we can say that it is more important to know how to count to 20 than it is to know what number comes after 14. But there is a difference between acknowledging that there are kinds of understanding that are culturally privileged in school mathematics, like knowing how to count to 20, and accepting that there may also be different mathematics at stake in our sensory politics. In terms of the latter, Healy and Fernandes (2010; 2011) explore how tactile means of learning about three-dimensional shapes enables blind students to learn mathematics, and they show how that which is taken to be a mathematical abstraction might be very different when it is expressed by means of moving hands. If abstraction is taken to be a process that depends on moving from visual perceptions of shapes to succinct verbal descriptions (listing the necessary properties of the shapes, for example), then there is no amount of moving with the hands that will count as an abstraction. But if abstraction can also be taken as the creation of an iconic gesture that expresses, for example, the way in which a hand will move when it touches the tip of a pyramid and runs its fingers along the edges until it reaches the base, then we begin to move towards a more expansive and inclusive approach to mathematical ability in the classroom.

The supposed disconnect between mathematical ability and the material world has come to inform certain assumptions about autism and Asperger's Syndrome. The stakes are high given the significant rise in the incidence of autism and the various political debates around diagnosis. Our aim here is not to enter into these debates, but to point out how mathematics has been taken up in the research on autism, sometimes in seemingly contradictory ways. In Asperger's seminal 1944 paper on autistic children, he claimed a link between mathematical ability and some of the children for whom 'abstraction is so highly developed that the relationship to the concrete, to objects and to people has been largely lost' (p. 85). Borrowing a refrain from the Romantics, Asperger wrote that 'a dash of autism is essential' in science or art, because the necessary ingredient for success 'may be an ability to turn away from the everyday world' (in James, 2003, p. 63). He also claimed that people with autism seem to speak without any apparent addressee,[13] 'as if into empty space,' and related this to their fleeting gaze, suggesting that the 'blinkered attitude' of autists makes them well-suited to mathematics or other careers that might require an almost solipsistic focus. Some researchers have suggested that the apparent absence or reduced empathy of autists, and the fact that far more male than female children are diagnosed with the condition, might explain the gender bias in advanced mathematics (Baron-Cohen, 2001). More recently, the psychologist Michael Fitzgerald has even raised the question of whether the cognitive style of people with Asperger Syndrome is also a 'mathematical style'. It thus seems that while eighteenth-century scholars fetishized blindness as a correlate of mathematical ability, the twenty-first century has opted for autism. These sorts of assertions about bodies and mathematical abilities reflect particular cultural concerns, often mythologizing mathematical ability while demoting the role of the human body in learning mathematics. The way in which former professor of mathematics Ioan James (2002, 2004) retroactively diagnoses legendary mathematicians such as Cauchy, Erdös, Galois, Gauss, Hardy, Lagrange, Lobachevsky and Riemann with forms of autism does not help us study the political framing of (dis)ability in mathematics.

Since Kanner (1943) offered the initial definition of autism (in terms of aloneness, language abnormalities and obsessive desire for maintaining of sameness, amongst other things), the general theoretical approach has been to make a strong distinction between children's relations to the material world and their relations to the social world. This approach aligns with the dominant and persistent developmental psychology approach that suggests autists suffer 'severe difficulties in the socio-cultural realm of other people' (Williams & Costall, 2001, p. 101). The dualism of the social and the material in much of the research on autism has cast autists as social misfits with limited capacity for emotional relationships. Furthermore, a developmental psychology approach decodes autists' interaction with objects in terms of their inability to 'pretend play', and it judges this in terms of

[13] This is much like mathematical writing, as Pimm and Sinclair (2009) discuss.

their limited grasp of meta-representation.[14] Meta-representation is said to include using an object in one of the following ways: 'Treating it as if it was something else, attributing properties to it which it does not have, or pretending it is present (i.e., gesturally) when in fact it is not' (p. 103). But the very idea of meta-representation that informs this judgement is based on assumptions about both matter and reasoning – assumptions that inclusive materialism contests.

Williams and Costall show how most research on autism has interpreted autistic interaction with material objects in terms of a strict ontological binary divide between objects and people, usually ascribing to a 'theory of mind' that correlates autism with a social affective disorder. Such a correlation is entirely disrupted when we shift the theoretical terrain and attend to 'social' relationships in terms of complex, entangled materialities.[15] Rethinking the agency of the material world and, indeed, rethinking affect as distributed across ever-growing material assemblages of rhythmic sensation, demands a radically new approach to understanding autism. A theory of sensation that resists the Kantian image of intuition and follows instead a more dispersed flow of affect and sensation could have huge impact. Some autists, for instance, describe their avoidance of another person's eyes in terms of the intensity of the encounter and the superabundance of face-affect that they feel on their own face (Manning & Massumi, 2010). Some austists claim that the habit of rhythmically moving their arms through the air is a way of feeling themselves as differentiated from the environment.[16] Some suggest that when they enter a room, they do not instantly decode it in terms of its objects, but instead perceive only provisional indeterminate objects, whose functions are not yet scripted (Manning & Massumi, 2010). Might this sensitivity to the unbounded nature of objects – in other words, an *ability* to perceive the provisionality of the assemblages and a diversity of affordances – allow them to engage with the materiality of a mathematical object and explore its potential 'mis-use' or its virtual indeterminacy? Such

[14] Moving away from this approach, Loveland (1991) has relied on the psychologist James Gibson's theories of affordances to rethink autists' interaction with objects. Although applications of Gibson's theories are very promising, many such studies analyze object-interaction in terms of anthropomorphic-centric use-value without adequate analysis of the interpenetration between the social and the material.

[15] Instead of dismissing attention to objects as being a deficit in abstraction skills, one might begin to think differently about autistic preference for exploring objects with proximal senses of touch, taste, smell and sound (Wing, 1969; Hermelin and O'Connor, 1970, cited in Williams & Costall, 2001).

[16] This can be seen in the Amanda Baggs video clip that we discuss in Chapter 5.

questions are meant to be provocative, in that they challenge us to rethink what it might mean to possess ability in mathematics.

In the next section, we draw on contemporary work in disability studies and continue to develop our notion of the body-as-assemblage, which was introduced in Chapter 1. In particular, we want to examine how competing models of (dis)ability attend to both the body's sociocultural nature and the real material conditions of embodiment, with all its deviations and excesses. We will do this by first tracing a brief history of (dis)ability studies and its changing, political models for thinking about and researching (dis) ability. We then discuss particular developments in (dis)ability studies that have begun to study the indeterministic nature of the human body, much like the post-humanist reconception of perception that we described earlier in this chapter. The exciting new work of James Overboe and others in this area offers a new way of thinking about perception that decentres human agency, thus furthering the post-humanist project. This work points to how learning assemblages might be conceived in terms of flows of affect and percept, rather than as receivers of particular organ-bound sensations or deployers of particular neurologically defined capacities. To wit, the awkward but compelling notion of an 'impersonal life' invites us to consider the shared dispersal of sensation across a body. There is much more research to be done in this area, and we are only able to offer a glimpse into all that it promises.

Reconfiguring the human body

The work of Michel Foucault in the 1970s on disrupting the medical gaze and overturning the entrapment of the body as a closed set of diseased organs was hugely influential in rethinking disability. This work was meant to reclaim the disabled body from dehumanizing medical and biological discourses; it was an attempt to wrestle the 'aberrant body' away from the medical establishment where its deviance was considered a deficit through various measures.

> Body theory helped to re-present the aberrant body as a contested field of political meanings, and deviance became defined as a subjective imposition upon neutral, divergently functioning organisms. (Snyder & Mitchell, 2001, p. 370)

Thus the 'social model of disability' was extremely important in political advocacy for those disenfranchised through exclusionary practices of segregation. It refrained from classifying and formulating the embodied experiences of the disabled, a practice which had been very debilitating, and turned its attention to analyzing the sociocultural conditions of (dis) ability.

> This neglect was wilful and strategic: it explicitly sought to leave an over-analysed entity mercifully alone. Instead, the critical gaze was turned on the practitioners of physical, occupational, and psychological evaluations and on the institutions that authorized their gaze. (Snyder & Mitchell, 2001, p. 374)

This social model was crucial in politicizing disability studies. As Shakespeare and Watson (2001) explain, distinguishing the physical impairments that people have from the way in which they are disabled by them in their social activities allows the gaze to be turned back on the institutional practices. The *Union of the Physically Impaired Against Segregation* (UPIAS) (1976/1997) describes the social model of (dis)ability in these terms:

> In our view, it is society which disables physically impaired people. Disability is something imposed on top of our impairments by the way we are unnecessarily isolated and excluded from full participation in society. Disabled people are therefore an oppressed group in society. To understand this it is necessary to grasp the distinction between the physical impairment and the social situation, called 'disability', of people with such impairment. Thus we define impairment as lacking all or part of a limb, or having a defective limb, organism or mechanism of the body and disability as the disadvantage or restriction of activity caused by a contemporary social organisation which takes little or no account of people who have physical impairments and thus excludes them from participation in the mainstream of social activities. (pp. 4–5)

But Shakespeare and Watson (2001) argue that the distinction between 'impairment' and 'disability', although intended to help redress social and economic inequity and to point to the social construction of disability, continues to support an ontology of the body that implicitly undermines attempts at political revision. Rather than debating which term to use in referring to (dis)abled bodies, they argue that the whole concept of physical normality needs to be problematized and that 'we have to recognise that disablement [impairment] is not merely the physical state of a small minority of people. *It is the normal condition of* humanity' (p. 18, *emphasis in original*). They argue that this 'ubiquity of impairment' is an empirical fact, although not one with which we regularly reckon. Obviously, such an assertion entails all sorts of problematic consequences – starting with the concern that some people clearly suffer more than others, and one needs a means of addressing this difference. Nonetheless, we wonder if this approach can help us get beyond the us/them thinking that haunts our attempts at 'special education'. We will return to this proposal later, but first we touch on some recent developments of the 'social model'.

Concerned that this 'social model' had abandoned all attempts to grapple with the lived embodiment of disability, there was a call in the 1990s for more phenomenological studies of (dis)ability, studies that would indeed address the visceral and tactile experiences of the disabled (Paterson & Hughes, 1999; Linton, 1998). The able body was no longer considered an ideal against which all others are measured, but rather an *aesthetic* product of cultural forces. By breaking up the alignment between disability and illness, where illness followed a predictable course, disability was redefined as '[t]hat in the body which exceeds deterministic efforts to predict a life trajectory' (Snyder & Mitchell, 2001, p. 377). In redirecting our attention to 'that in a body' that breaks with deterministic images of life's trajectory, this approach to disability points to the way bodies are always caught up in processes of becoming – processes that cannot be prescribed by genetic code – despite the comfort we might take in looking for such definitive explanations. This approach opens up the future, demanding that we look at life – just as we look at genes – as fuelled by chance and unscripted recombination. Focusing on the chance-inflected process of inheritance might allow us to imagine becoming radically unrecognizable to our previous embodied selves.

The neurodiversity movement also shifted the political terrain by advocating for the right to cognitive difference while resisting others' claims that such differences need to be corrected. Recent research on dyslexia, for instance, celebrates the exceptional spatial reasoning of dyslexics rather than dwelling on the difficulties they experience with the alphanumeric (Eide & Eide, 2011). Kalvis Jansons, a dyslexic mathematician and professor at University College London, describes how his penchant for visualization and diagrams is deeply bound up with other sensory modalities: 'It would be a mistake to believe … that non-verbal [spatial] reasoning has to involve pictures. For example, three-dimensional space can be equally well represented in what I often think of as a tactile world' (quoted in Eide & Eide, 2011, p. 59). Dyslexics may struggle with procedural learning and rote memory tasks, but their memory of phenomenological details – that is, details pertaining to physical aspects of an experience, such as tactile, motor or spatial arrangements – exceed that of non-dyslexics (Eide & Eide, 2011). A school mathematics that demotes the visual-haptic-spatial aspects of sense-making in mathematics creates a disadvantage to a large population of students.[17] The high percentage of successful engineers, architects and artists with dyslexia indicates how much potential is probably going untapped in *school* mathematics (West, 2004). Another form of this untapped potential is described by Mitchell (2001), who writes about 'seeing disability', by which he means that the sighted do not see the 'dangerous displacements, removals, and disappearances' that the blind are able to perceive (pp. 392–393). Disability becomes ability, and vice versa.

Echoing the proposal of Shakespeare and Watson (2001), we put forward here the political project of 'becoming imperceptible', or becoming impaired – a means of problematizing 'ablism' – so that we might begin to understand the ways that our learning assemblages are already always dysfunctional or disabled. As Mitchell (2001) suggests, disability studies shows us how to problematize the very idea of mathematical ability as a singular achievement by an individual:

> To recognize this is not to deny the very real difference that actual physical disabilities make in the lives of individuals, nor to question the urgency of concrete political and ethical action on behalf of the disabled. It is rather to question the whole model of self-sufficiency, mastery and perfect ability (not to mention perfectibility) that governs our image of the individual. (p. 396)

The sociologist and disability theorist James Overboe has done just this, questioning the model of the *able individual*, by arguing that the construct of an 'impersonal life' might allow for a radical politics of (dis)ability. Overboe (2009a, 2009b) shows how his own experiences with cerebral palsy should be seen in terms of a life of pure immanence, a life conceived in terms of 'post-person' vitalism sustained through bodily expression, rather than humanist conceptions of the self. Instead of pursuing the self-affirming narratives promoted by those engaging in phenomenological studies of (dis)ability, Overboe (2007a; 2007b; 2007c) wants to avoid the

[17] Current data indicates that 10–15% of the U.S. population are labelled as dyslexic. These rates seem to be lower in other countries. In Canada, about 3.2% of school-age children are described as having 'learning limitations', which include dyslexia. European statistics indicate somewhere between 2 and 3% of the population are dyslexic. Of course, different statistics take different definitions of dyslexia. About 60% of students diagnosed as ADHD are also diagnosed with dyslexia, so the disadvantage that is built into the curriculum becomes increasingly evident. Moreover, many of the spatial tasks in typical IQ tests, as well as those found in standard state tests, focus on two-dimensional Euclidean spatial relationships rather than three-dimensional spatial reasoning skills, like those that are most often demonstrated by dyslexics.

way in which such narratives tacitly affirm a limited view of life as that which is grounded in a wilful and rational subject. His aim is to think past (dis)ability as something that must be overcome, to move beyond a 'docile body' whose identity must be recovered and celebrated, because such approaches will systematically 'invalidate the lived experience' of those who cannot meet such expectations' (Overboe, 2009a, p. 245). Instead, Overboe pursues the political agenda of addressing injustice and inequity, but he does so through post-humanist ideas about the body. Rather than valorizing the wilful agentic subject who overcomes a disability, Overboe's approach argues that disability studies must find a way to think about the impersonal nature of dispersed 'agency' without recovering or recuperating the humanist subject. The humanist subject offers a 'limited view of life [that] fails to affirm disabled lives that are simply expressed without cognition, intent or agency' (p. 243).

The notion of 'impersonal life' is a challenging one, as it suggests going not just without person, but beyond person. Overboe uses the term to push us into considerations of a body that does not belong to us in any definitive way – it is *not* personal. An impersonal life is a life that is networked and entangled in ways that trouble the conventional conception of person. This impersonal life is a 'pre-reflexive and impersonal consciousness, a qualitative duration of consciousness without a self ...' (Deleuze, 2001, p. 25). We suggest here, following Overboe, that attending to the intensity of affect – the impersonal 'communication' or expression that singularizes a sensation or a perception – helps us move away from an identity politics and towards a micropolitics of pre-individual singularities, towards potentialities that have yet to be imagined. Goodley (2009) makes a similar argument, suggesting that 'the metaphor of the body as organism refers to an endemic societal view of the body as sovereign self: contained, knowable, measurable, and dis/abled' (p. 264). We believe this is an extremely important argument to pursue, as it aims to reckon with the way that sensation is dispersed across a situation with a provisional set of organs open to constant reorganisation. The movement of sensation across an entangled system sustains life – that is, 'impersonal life' – as that which is shared and not individual.

Of course, the body needs to be recognized as an organism at particular times, and it is regularly hailed as such by medical discourse and often sustained as such through medical treatment. This we are not denying. But focus on the body as contained, knowable, measurable and (dis)abled can also work against the realization of equitable practices. In schools, the (dis)abled body is labelled as an isolated, bound and known organism, and although identification of a disability indisputably benefits many children, the practice as implemented often assumes that the student's body is a self-contained (but impaired), individuated entity. Our proposal works against common institutional practices that are well-entrenched and relate to the needs of governmentality: The body is not welcome in classrooms unless it is known to be bounded. As Goodley (2009) suggests, 'educational institutions and discourses require (children's) bodies to be measurable, accessible, easily treatable and understood. Bodies that morph in ways that are unintelligible – that are not dis/abled enough or fit the typology of dis/abled used by educational institutions – are not easily included' (p. 264). A post-humanist approach moves beyond the 'accommodations' from which an individual might benefit and begins to conceptualize the body as collective and unknown. Perhaps, as Goodley remarks, 'disabled people are dismodernist subjects *par excellence* in their engagements with technology and science' (p. 267). Instead of the body as organism, one might begin to embrace the body with(out) organs in some fundamental way; 'the body-with-and-without-organs is a blurred metaphor that captures the hybridized and cyborg-like nature of disabled bodies' (p. 267).

This approach to education might be described as a 'posthumanist pedagogy', demanding that we move beyond the cybernetic images of a cyborgian body and begin to engage with learning assemblages as flows of affect and percept (Hickey-Moody, 2009). The concept of flow problematizes traditional ideas about agency. Of course, this is precisely the aim of Barad's work, in which the notion of intra-action is meant to displace agency from particular fixed objects (be they people, tools or concepts). Ingold (2008) more forcefully articulates a position against the fascination with agency when he writes: 'Indeed, the more that theorists have to say about agency, the less they seem to have to say about life; I would like to put this emphasis in reverse' (p. 211). He notes that the term 'agency' will always be burdened by the assumption that there is a unitary source for action – that being the agent. Like Ingold, we embrace a post-humanist approach to the body, and we focus on the flows and perturbations of materiality not only as a study of body mechanics – or individualized agency – but as an attempt to learn more about the life-processes entailed in teaching and learning mathematics. Ingold is concerned that a theoretical stance that centres on materiality will focus on the agency of objects rather than the flow of life forces. Like other post-humanist theorists, he is grappling for the appropriate language to study the *aliveness* of an indeterminate world. He suggests that life be studied not for how it is captured or contained in bodies and objects (and their respective or joined agencies), but for how these forms discharge, leak and flow.

Like Ingold, we aim to follow the flow and variation of materiality always in movement and time. Rather than dismiss this as a dangerous move towards passivity – a move that might undermine political action on behalf of those who suffer from a (dis)ability – we think this approach might allow educators to study 'ability' as a collective capacity of force, or an entanglement of affect. Affectivity of this kind cannot be reduced to a causal mechanism, seeing as it involves the problematic production of an excess of, or something over and above, the network of corporeal causes. Learning will always unpredictably outstretch the determinate actions taken to effect learning.

In this chapter we have focused on various aspects of mathematical sense, with a particular emphasis on the sensory organs. In the next chapter, we continue our exploration of mathematical sense but with an emphasis on how different ways of sensing the world come to align with – but sometimes break from – what is taken to be 'common sense'. We focus on the historical development of a *mathematic aesthetic* and show how this aesthetic is taken up within mathematics education.

7 Mapping the cultural formation of the mathematical aesthetic

In this chapter, we explore some political dimensions of the mathematical aesthetic. We draw on the work of Jacques Rancière to argue that the mathematical aesthetic must be analysed as a form of cultural politics. We claim that aesthetic practices are political practices precisely because they partake in 'the distribution of the sensible', a distribution that discloses and determines that which is held in common within a particular community (Rancière, 2004). Aesthetic practices are simply ways of 'doing and making' among many other ways of doing and making, but they are uniquely generative of forms of visibility and sensibility, and are thereby central to determining what others might call membership in a community of practice. As Rancière argues, aesthetic practices operate through a paradoxical mix of *autonomy* and *dependence*, on the one hand free from the demands of functionality and explanation (a painting is only a painting if it is *not* useful), and on the other hand entirely reliant on sensory effect (a painting is only a painting if it is perceived). We argue in this chapter that a mathematical aesthetic operates through the same paradoxical mix. Our aim is to show how these aesthetic practices function in mathematics, as well as in mathematics education, and to indicate that any particular distribution of the sensible privileges certain forms of sensibilities over others. In other words, any particular drawing of the boundary between what makes mathematical sense and what does not entails a particular kind of consensus about the valuing and regulating of the senses. We will draw on our exploration of the mathematical aesthetic to argue that instead of seeking new – and possibly more invasive – forms of consensus-making in mathematics education, we might instead look for ways of perturbing current aesthetic regimes. Rancière's notions of *consensus* and *dissensus* enable us to explore current and alternate ways of making sense in mathematics classrooms. To illustrate these important notions, we begin with a short example from the mathematics classroom where both consensus and dissensus can be seen in action.

Classroom consensus and dissensus

Mathematics classroom discourse often aims for consensus, not simply through an emphasis on one correct answer, but also through the primacy of the alphanumeric as the ultimate form of communication. Spatial sense is typically downplayed in classroom activity, while haptic and other senses are rarely, if ever, considered. And yet mathematical consensus marshals all of the senses to make itself felt. And with equal force, the senses are the site of resistance to that same consensus. Consider, for instance, a well-known video excerpt of a grade three classroom,[1] in which a student named Sean suggests that six is both odd and even because three is an odd number and six is three groups of two. This video excerpt has been discussed and analysed at length in the mathematics education literature.[2] Ball (1993) uses it to examine the dilemma of respecting children as mathematical thinkers. Sinclair (2010) interprets the episode in terms of the contrasting intellectual passions to which Sean and his classmate Mei are committed. Here, we propose to re-read the episode with an eye to the political, that is, in terms of the acts that create and/or disrupt normative ways of making sense of mathematics in the classroom.

The class has been working with patterns involving even and odd numbers when, one day, Sean announces that he had been thinking that 'six could be both odd and even' because it is made of 'three twos'. The students discuss his proposal and dispute its legitimacy. Sean notes that not all even numbers are also odd, but that 6 and 10 are because they can be considered odd or 'unfair' groupings. He valiantly defends his assertion in the face of a growing concern on behalf of the other students. Another student – Cassandra – disagrees with Sean and goes to the board to show why, picking up the pointer and reaching up to point at the visible number line above the board. She rhythmically counts off the numbers 'even, odd, even, odd, even, odd ...', as though the physical act of repeatedly banging the pointer against the number line shows why six cannot be odd. One can see in her action a rhythmic and ritual enactment of the autonomy of the even-odd number pattern. She is literally performing how the pattern has a certain automatic unfolding logic in it. One can also see in her action the way in which the body is

[1] This video is available at http://ummedia04.rs.itd.umich.edu/~dams/umgeneral/seannumbers-ofala-xy_subtitled_59110_QuickTimeLarge.mov.

[2] In earlier reports of the data, the boy was referred to as Shea.

implicated in the performance of this autonomy. We see here an important gesture that aligns sense (as sensation) with sense (as meaning). Sean refuses to accept this binary, nor is he persuaded by the temporal enactment of odds and evens, and he returns to his partitioning-based argument.

At this point in the lesson, the teacher makes a move to reintroduce 'common sense' by asking the class to give a show of hands indicating who knows the 'working definition' of even and odd. The show of hands is another embodied gesture that makes visible the commonality of common sense, especially in this case, as it physically demonstrates a shared commitment to a definition of the mathematical concept under discussion. The teacher, after listening as Sean offers his argument another time, draws six circles on the blackboard while asking 'are you saying that all numbers are odd then?' Sean uses these circles, dividing them into three groups of two, to 'prove' to his classmates that six is also odd. The class thus moves away from the rhythmic tapping on the number line as the material terrain for establishing parity/disparity towards the discrete object-driven view of number, where each number stands on its own, individuated by property rather than by sequence. The flow of conversation is so seamless that the major ontological disruption – from a temporal, alternating definition of odd/even to a differently structured one – passes by unnoticed.

After working with other examples of even and odd numbers (10 and 21), also involving the partitioning of circles, another student (Mei), who also disagrees with Sean, sums up the concern provoked by his disruptive act, stating: '[L]ike if you keep on going like that and you say that other numbers are odd and even, maybe we'll end it up with all numbers are odd and even. Then it won't make sense that all numbers should be odd and even, because if all numbers were odd and even, we wouldn't be even having this discussion!' Indeed, it is Mei's vision of what 'make[s] sense' that aligns with the conventional mathematical definitions of even and odd, and her eloquent argument has on more than one occasion led viewers (of the video) to comment on her mathematical sophistication.

In contrast to Mei, Sean's contribution breaks with common number sense and offers an alternative way of organising the natural numbers in terms of factors.[3] Sean justifies his suggestion that some numbers are

[3] Indeed, Sean is indirectly distinguishing even numbers that contain at least one odd factor from those that do not, the latter having the specific label of 'powers of two' in mathematics. Sean's 'even-and-odd' numbers are all the even numbers that are not powers of two. He is gesturing towards a tripartite division of numbers that consists of the totally odd, the odd-and-even, and the totally even. Such a categorisation is entirely mathematically defensible and indeed, in particular problem situations, perfectly functional.

both odd and even by showing how these numbers are *more* than even. In other words, one might focus on how numbers like six can be generated as a set consisting of *an odd number of things*. He offers a new way of partitioning these numbers and disrupts the binary logic of even-or-odd. In pursuing this new partition of numbers, Sean troubles the current way of making sense in the classroom.

We see Sean's contribution as an act of dissensus, in that it makes visible and audible what was previously invisible and inaudible; dissensus, as Rancière (2004) proposes, 'enacts a different *sharing* of the sensible' (p. 7). Dissensus is often a short-lived moment of dispute when the distribution of the sensible is contested, when someone stands, speaks out, touches an untouchable, eats a forbidden fruit or gazes into a once-veiled object, a moment when the senses are used 'improperly' to dispute the equality of common sense. If consensus is an alignment between sense (as sensation) and sense (as meaning), then we use the term dissensus to refer to that which breaks up this alignment. Rancière (2009) suggests that a political and polemical redistribution of the senses – a 'dissensual supplement to the other forms of human gathering' (p. 32) – offers us a place to start thinking differently about community. An act of dissensus is a controversial disruption of the limits of the sensible in any given collective situation. Acts of dissensus operate on the ragged boundary between the aesthetic and the non-aesthetic; that is, they often operate as sites of non-sense, where sense is dislocated from meaning. They are border crossings, shredding the borders and divisions that currently partition the sensible. In Sean's alternative partitioning of the natural numbers, we see how dissensus is not an overturning of institutions: It does not simply reorder hierarchies of power. Dissensus is:

> [a]n activity that cuts across forms of cultural and identity belonging and hierarchies between discourses and genres, working to introduce new subjects and heterogeneous objects into the field of perception. (Rancière, 2010, p. 2)

Thus, acts of dissensus introduce new subjects and objects into the field of perception. In the previous analysis, we focused on one particular new object – a number that is both even and odd – but Sean is also a newly configured subject who is newly entangled in the concepts he perceives. The subject comes into being through both consensus (alignment with common sense) and through dissensus (divergent individuation). The senses become sites where subjects exhibiting difference and diversity are either recognized as intelligible (visible, audible, etc.) or unintelligible (invisible, inaudible, etc.).

Following this line of argument, one can unpack the term 'community of practice' in terms of *community of sense* so as to describe how cultural politics maps onto classrooms through the senses. In classrooms, the sensible is distributed and partitioned into forms that fuse visibility (or audibility, etc.) with intelligibility. By focusing on the role of the senses in delineating membership in a community of practice, we can begin to study the contingency of intelligibility to show how sense-making might be done differently. Rancière's (2009) community of sense is not about agreed-upon ways of doing things in the classroom, as in Yackel and Cobb's (1996) 'sociomathematical norms', which function 'above the senses' in that they focus almost exclusively on discourse. Rather, Rancière takes a more materialist approach by focusing on the partitioning of the sensible:

> I do not take the phrase 'community of sense' to mean a collectivity shaped by some common feeling. I understand it as a frame of visibility and intelligibility that puts things or practices together under the same meaning, which shapes thereby a certain sense of community. A community of sense is a certain cutting out of space and time that binds together practices, forms of visibility, and patterns of intelligibility. I call this cutting out and this linkage a partition of the sensible. (Rancière, 2009, p. 31)

His correlation between the senses and intelligibility offers subtle but significant insight into the way that meaning-making emerges in classrooms. The term 'intelligibility' is here used to point to the fusing of the 'true' with the 'sensible' in what is taken as *common* to the community. As we discuss later in this chapter, this correlation has influenced the kinds of mathematical practices that have become valued in policy and curriculum. We first discuss, however, the novel way in which Rancière formulates the intersections of politics and aesthetics, and then we extend this discussion to the mathematical aesthetic in order to examine how autonomy works in mathematics.

A political aesthetics

In his *Letters on the aesthetic education of mankind*, the eighteenth-century poet and philosopher Friedrich Schiller put forward a theory of aesthetics that aimed to bear 'the whole edifice of the art of the beautiful and of the still more difficult art of living' (quoted in Rancière, 2010, p. 115). It is this paradoxical coupling of art and life under the banner of aesthetics, suggests Rancière, that has made aesthetics so enigmatic and difficult to study.

And yet it is precisely this odd coupling that has allowed the aesthetic to function so effectively as part of the political fabric of life. The aesthetic we have inherited from this tradition is radically different from other ways of embracing art, such as those that construe art as a representation or copy of the true, or those that construe art as a shaping of matter, for each of these entails an ontological divide between *the art of the beautiful* and *the art of living*. According to Rancière, the Western tradition of aesthetics confounds two oppositional concepts of sense, the first associated with the autonomy of art and the second with collective or common forms of sensibility. Thus, the aesthetic operates through the conjunction of pure 'sense' and common sense, conditioning our modes of individual perception, as well as our social institutions. In other words, political participation and 'artistic' practices are reciprocally implicated, not simply in terms of class and *judgements* of taste, but in terms of the material distribution of what is taken to be *sensible*.

> The entire question of the 'politics of aesthetics' – in other words, of the aesthetic regime of art – turns on this short conjunction. The aesthetic experience is effective inasmuch as it is the experience of that *and*. (Rancière, 2002, p. 134 *emphasis in original*)

But it is not simply art qua art that partakes of the aesthetic experience, for one can study the tension of this awkward conjunction in other activities. In mathematics, for instance, one might think of this 'and' in terms of the complex ways in which form and function are mutually entailed in our sensory experiences: the way, for instance, that we attend to the formal qualities of a cube (edges, faces, angles) *and* simultaneously perceive its affordances and capacities for movement or activity (as when we work on a Rubik's cube). Or one might think of this conjunction in terms of the singularity of sensory encounters (the way sensory experiences individualize the body) *and* the making of consensus through the valuing and regulating of the senses. The eye, for instance, perceives the cube as a distinctive relationship but is simultaneously calibrated to see certain things through the forms of attending encouraged in the classroom. This powerful enfolding of politics with aesthetics operates through the senses. Decoding a cube in terms of the agreed-upon markings that designate congruence entails a falling-into-step of the senses and their alignment with common sense. Similarly, classroom tasks that direct students to correlate equations with tables of numbers determine the limits of the sensible by presenting mathematics in such forms – these tables and equations become the surface of mathematics, beyond which we cannot reach. At the same time, these tasks stage the 'proper' forms of perception that are appropriate for this community of practice.

The mathematical aesthetic, according to Rancière's use of *aesthetic*, thus acts to conjoin the purity of logical deduction with embodied sensory engagement. This can be seen beautifully in the videotapes of a mathematics lecturer discussed by Núñez (2006), in which the lecturer offers a highly formal verbal description of infinity, while at the same time performing infinity by rhythmically gesturing a repeated, linear movement away from his body. While his speech negates his own being – through the detemporalized, decontextualized and depersonalised voice of the mathematical discourse – his hand insists on carving out the iterative space. But the sensory engagement involves affect, as well. As Lockhart (2009) suggests in his popular essay, *A mathematician's lament*, a mathematical argument 'should feel like a splash of cool water, and be a beacon of light – it should refresh the spirit and illuminate the mind' (p. 68). The demand that mathematics touch our being like 'a splash of cool water', as though it were meant both to wake us up and to cleanse us, reveals how it is meant to operate through this aesthetic duality, this conjunction of enlightenment (purity) and refreshment. Rancière (2002) points to how the aesthetic regime operates in cultural formations more generally, showing how it functions as a policing force in 'totalitarian attempts at making the community into a work of art', as well as in 'the everyday aestheticized life of a liberal society and its commercial entertainment' (p. 133). Thus, aesthetics operates by folding the sensory fabric of the common, which separates inside from outside, partitioning both the social and material worlds, while entwining or coupling what is 'in' and what is 'out', so that no line of inventive flight is ever entirely free.

At work in all of this is a twisted concept of autonomy, one that grants independence to the aesthetic object and then takes it away. In other words, the duality of the aesthetic object demands, on the one hand, autonomy from functionality and explanation, while on the other hand, it emerges through a dependence on the everyday sensory modalities of embodiment. From the modernist tradition of aesthetics, we inherit an image of the art object as autonomous and its aesthetic qualities as indefinable and out of reach. Ironically, art is not a *work* of art (compare a painting by Jackson Pollock and a picture of Pollock painting in his signature style), because the aesthetic qualities are meant to be somehow contained within the art object. In other words, any labour involved in the production of art is deemed tangential to its power and presence. One can see in this approach a similarity to popular conceptions of mathematics – which we explore in more detail in the next section – where the mathematician is cast as someone whose insight or intuition has led to the discovery of a mathematical object, rather than someone who has laboured to produce a mathematical object.

Any discussion of the aesthetic element in mathematics has to grapple with the way aesthetic practices are conceived in relation to this concept of autonomy. An activity is considered art insofar as it partakes of this autonomy and, paradoxically, disconnects itself from its own making – 'art is art to the extent that it is something else than art' (Rancière, 2002, p. 137). One can find this paradox at work in most theories of the aesthetic which rely on a mystical preconscious source for inspiration. For instance, Hegel argued that the artist can never entirely know the source of his or her expression, for it springs from the unconscious; similarly, in 1908 Poincaré asserted the unconscious sources of mathematical discernment. When mathematics is seen as driven primarily by aesthetic principles, as we discuss later in this chapter, the mathematician is subject to a certain *subjective anonymity* as his or her conscious presence is displaced. According to Rancière, this paradox of aesthetic sense – a lived paradox in which the autonomy and separateness of the aesthetic sense is opposed to the aspiration to *live* it as a sensibility – is actually the source of its political power. The aesthetic operates through the dream of an unavailable ideal form that must be made flesh and possessed as reality. And although we are concerned here with how this maps onto experiences of mathematics teaching and learning, it is important first to understand the ways in which this complex tradition of aesthetics has come to figure more generally in our everyday lives. One could argue, for instance, that affluent segments of Western society now look at the world and their community as though they were forms of art. Through ornament, industrial design and fashion, life – for some – has become increasingly aestheticized. To live life 'properly' in contemporary Western society is to appreciate the aesthetic qualities of one's environment and to produce or communicate in ways that reveal one's capacity for 'free' expression. Aesthetics removes art from the world through its autonomy but then inserts it back into the material world and demands that we live this autonomy through acts of freedom. The aesthetic regime of art shuttles back and forth between two scenarios in which art and non-art are linked – each with its own vanishing point or point of collapse: in one, life consumes art, and in the other, art denies life.

The paradox of aesthetics seems to thrive in contemporary consumer culture. We continue to invest in modernism's partitioning of the perceptible, whereby the commodity becomes fetishized by means of its being perceived as an aesthetic object. Indeed, an emphasis on design, as opposed to consumption, fosters a social commitment to art as a mode of collective education, whereby the population may be educated – through the senses – in how to live an aesthetic life. This occurs in the everyday consumer culture by which we obtain sustenance, and also in how we collect what were once prosaic objects and value them as antiques. Education figures prominently in the operation of this aesthetic regime. Education is devoted to enlisting students into particular habits of sensing and to ensuring that they internalize these forms of common sense. We argue in the following section that the mathematical aesthetic must also be read in terms of this political plot, showing how the aesthetic regime of mathematics operates through this duality by first claiming that mathematics partakes of the autonomy of the aesthetic and then insisting that one must live this aesthetic as a form of life.

The mathematician's sensibility

Instead of being concerned with actual senses (or a lack thereof), the literature on aesthetics and mathematics reverts to a mystification of the mathematician's sensibilities, as though the mathematician possessed some sort of intrinsic capacity to see or perceive (metaphorically, of course) either the immaterial entities of an ideal mathematical world or the masked mathematical structure of the physical world. According to this image, the mathematician is able to reconfigure the fabric of sensory life because he or she *perceives* what was previously invisible or insensible,[4] whether that be the rules governing some sort of physical process or the behaviour of a sequence of numbers. Rather than simply pointing out how this literature is a self-legitimizing discourse, our approach aims to unravel the specific means by which the aesthetic regime of mathematics ensures the visibility of mathematical objects and makes them available to thought. Lockhart's (2009) description of mathematics as an art form, and in particular the idea of inserting a line into a diagram, captures this common discourse in the literature concerning mystery and visibility:

> Now where did this idea of mine come from? How did I know to draw that line? How does a painter know where to put his brush? Inspiration, experience, trial and error, dumb luck. That's the art of it, creating these

[4] As we discuss in Chapter 6, the metaphorical aspect of this mathematical perception reached a high point in the eighteenth century, when it was considered by some to be advantageous for a mathematician to be blind. The blind mathematician would have unfettered access to the ideal, abstract objects of mathematics, including geometrical ones. These philosophers also argued that a lack of sight might compromise the mathematician's openness to the world around them, thus diminishing their moral sensitivity. The interplay between sense and common sense here achieves a staggering and polarised simplicity.

> beautiful little poems of thought, these sonnets of pure reason. There is something so wonderfully transformational about this art form. The relationship between the triangle and the rectangle was a mystery, and then that one little line made it obvious. I couldn't see, and then all of a sudden I could. Somehow, I was able to create a profound simple beauty out of nothing, and change myself in the process. Isn't that what art is all about? (p. 27)

The notion of pattern perception has emerged as a dominant mode of sensing in mathematics. For example, in his attempt to define mathematics as the 'classification and study of all possible patterns' (p. 12), the mathematician W. W. Sawyer (1955) implies that the heuristic value of mathematical beauty stems from mathematicians' sensitivity to pattern and originates in their belief that '*where there is pattern there is significance*' (p. 36; *emphasis in original*). Sawyer goes on to explain the heuristic value of attending to pattern:

> If in a mathematical work of any kind we find that a certain striking pattern recurs, it is always suggested that we should investigate why it occurs. It is bound to have some meaning, which we can grasp as an idea rather than as a collection of symbols. (p. 36)

Sawyer's claims resonate with how Poincaré describes the mathematician's special aesthetic sensibility as a sensibility towards pattern, which is viewed broadly as any regularity that can be recognized by the mind. For Poincaré, the mathematician is not only able to recognize regularities and symmetries, but is also attuned to look for and respond to them with further investigation. In an essay published late in his career, Alfred North Whitehead revisits Plato's ethical, or perhaps political, disquisition on mathematics and suggests an even grander visibility offered by pattern. Whitehead seems bewitched by the axiological connection Plato made between mathematics and the good and, having admitted the incompleteness of Plato's argument, offers his own:

> We cannot understand the flux which constitutes our human experience unless we realize that it is raised above the futility of infinitude by various successive types of modes of emphasis which generate the active energy of a finite assemblage. The superstitious awe of infinitude has been the bane of philosophy. The infinite has no properties. All value has the gift of finitude which is the necessary condition for activity. Also activity means the origination of patterns of assemblage, and mathematics is the study of pattern. Here we find the essential clue which relates mathematics to the study of the good, and the study of the bad. (1951, p. 674)

For Whitehead, thinking about pattern is, at its core, an ethical activity, insofar as it concerns our experience, which is shot through with the insistence of finitude. Whitehead is much less concerned with promoting scientific advances (to which the utility of mathematics most often contributes) than he is with promoting creative, flexible and non-dualistic thinking about the 'human condition'. While Whitehead saw an affinity between mathematics and moral development, or at least what he took to be part of the 'human condition', most people who consider mathematics and morality in the same breath take a more cautious point of view (Hersh, 1997).

This emphasis on pattern, symmetry and regularity underscores the kind of autonomy that epitomises many accounts of the mathematical aesthetic. Mathematical patterns can be characterized as automata, in that they operate according to an intrinsic logic or rule, independent of outside stimulus or human intervention. No matter how much material force one can muster, one cannot disrupt or alter the unfolding of the pattern. A mathematician might 'explain' a pattern with reference to the actions or operations that might be used to produce it (adding, multiplying, tripling, etc.), but this activity or labour does not engender the pattern. Automata are 'self-acting', and patterns are self-engendering. The emphasis on *detecting* patterns demands that the mathematician perceive or sense that which is independent of his or her labour. The mathematician must 'grasp as an idea' that which is autonomous – that is, the mathematician must internalize the autonomy and *live* the mathematical aesthetic as a form of life. According to Rancière, this is precisely how an aesthetic regime operates – by insisting on the conjunction of two oppositional concepts of sense, the first associated with the autonomy of expression (or, in this case, pattern) and the second with the enactment of a common form of sensibility (in this case, the mathematician's). This impossible demand to *live* that which is autonomous is reflected in the contemporary literature on the mathematical aesthetic. As Lockhart (2009) suggests:

> To do mathematics is to engage in an act of discovery and conjecture, intuition and inspiration; to be in a state of confusion – not because it makes no sense to you, but because you *gave* it sense and you still don't understand what your creation is up to; to have a breakthrough idea; to be frustrated as an artist; to be awed and overwhelmed by an almost painful beauty; to be *alive*, damn it. (pp. 37–38, *italics in original*).

Poincaré (1908/1956), however, suggests that only the most creative mathematicians have access to an aesthetic sensibility. Similarly, Hardy (1940) claimed that 'the aesthetic appeal of mathematics may be very real for a chosen few' (p. 86). Any attempt to rethink the mathematical aesthetic in materialist terms has to reckon with this common inclination to assign a *selective* sensibility to the mathematician. Russell's (1919) famous quotation is frequently cited as an example of how this sensibility is expressed: 'Mathematics, rightly viewed, possesses not only truth, but supreme beauty – a beauty cold and austere, like that of sculpture' (p. 60). A cold and austere beauty is one that is pure and independent, or erases all traces, of the hands that made it. Moreover, the stone cold sculpture shows no interiority and no emotion, and the matter is hard and resistant to the form it is given. And yet, like the quotation from Russell, a focus on mathematicians' aesthetic considerations is frequently offered as a counter to the usual emphasis on purely epistemological concerns about truth and certainty. The historian of mathematics Morris Kline (1953), for instance, points out that aesthetic concerns not only guide the direction of an investigation, but also motivate the search for new proofs of theorems that were already correctly established but lacking in aesthetic appeal. Kline concludes that this aesthetic motivation is a definitive sign of the artistic nature of mathematics. The distinction between functionality and autonomy operates in this judgement, whereby a mathematical proof becomes aesthetic as it is granted a certain autonomy, that is, as it comes to be of no apparent everyday value. In other words, once we have a proof, what is the *use* of another? In not being useful, it is deemed aesthetic. Lockhart's (2009) lament for the loss of mathematics-as-art reveals a similar distinction, as he distinguishes the 'mundane "useful" aspects [that] would follow naturally as a trivial by-product' from the more central aesthetic activity of real mathematics. The latter is precisely how the mathematician Wolfgang Krull (1930/1987) characterizes mathematics, by contrasting epistemic concerns about truth and logical consistency with aesthetic engagement:

> Mathematicians are not concerned merely with finding and proving theorems, they also want to arrange and assemble the theorems so that they appear not only correct but evident and compelling. Such a goal, I feel, is aesthetic rather than epistemological. (p. 49)

The epistemic in terms of the determination of true or false is here aligned with the functionality of mathematics, whereas being 'evident and compelling' is aligned with the aesthetic. This distinction construes mathematics as both autonomous (evident) and affective (compelling) in its aesthetic dimension. Indeed, we are compelled to submit to mathematics only when it achieves this aesthetic dimension, for it is only then that it truly embraces its autonomy. Rota (1997) draws attention to the way in which aesthetic descriptors used by mathematicians may, in fact, represent veiled ways of sustaining an image of mathematics as immaterial and 'immune' to the sensual:

> Mathematical beauty is the expression mathematicians have invented in order to obliquely admit the phenomenon of enlightenment while avoiding acknowledgement of the fuzziness of this phenomenon. [...] This cop-out is one step in a cherished activity of mathematicians, that of building a perfect world immune to the messiness of the ordinary world, a world where what we think *should* be true turns out to *be* true, a world that is free from the disappointments, the ambiguities, the failures of that other world in which we live. (pp. 132–133; *emphasis in original*)

This tension between the mathematical aesthetic and the messiness of the ordinary world reflects Rancière's depiction of an aesthetics that partitions the sensible into either access or denial of access. According to Rota, mathematical beauty is a way of sustaining an untouchable or unreachable sensory realm, for it is through this aesthetic judgement that the ideal becomes real. The aesthetic grants mathematics a sensory aspect, while simultaneously denying access to this encounter for all but a few.

In 1940, G. H. Hardy published what arguably became the most widely read inquiry into the mathematical aesthetic. Hardy was primarily interested in defining mathematical beauty as it exists in mathematical products, particularly in proofs. He proposed a somewhat complex scheme that distinguished 'trivial' beauty – which can be found, for instance, in chess – from 'important' beauty – which can only be found in serious mathematics. Serious mathematics requires depth and generality (scope and reach) if it is to be significant, but because none of these aesthetic qualities can be defined, only those with a 'high degree of mathematical sophistication' (p. 103) can recognize them. Such mathematicians will find a mathematical idea significant when it can be 'connected, in a natural and illuminating way, with a large complex of other mathematical ideas' (p. 89). It is the clause 'natural and illuminating' that inscribes a common ('natural') form on a particular way of sensing and/or making intelligible ('illuminating').

Hardy illustrates his notion of mathematical beauty with two examples: Euclid's proof of the infinity of primes and the Pythagorean proof of the irrationality of $\sqrt{2}$. These two proofs appear frequently in the literature as particularly fine examples of beautiful proofs. Dreyfus and Eisenberg (1986) showed five different proofs of the claim that $\sqrt{2}$ is not rational (all using indirect reasoning) to a group of mathematicians who were asked to identify the proofs that were most elegant. All of them selected the same pair and justified their choice in terms of the perceived simplicity and minimal amount of background knowledge required to understand the proof. Both of these judgements – simplicity and minimal assumed knowledge – point to particular desires that fuel the mathematical aesthetic. On the one hand, simplicity is contrasted with unnecessary adornment and a preference for truth to be present and singular, without difference or complication. On the other hand, minimal assumed knowledge relates to the desire that mathematics ultimately concern pure reason rather than knowledge, because knowledge is always tainted with the particularities of its historical context. The tension revealed within these desires echoes the larger tension of aesthetics more generally. But what remains unexamined in the Dreyfus and Eisenberg study is the way that simplicity and minimal assumed knowledge are valued in relation to the *result* of the proof.

In other words, we need to explore further what is actually entailed in these two proofs. They are both proofs that deploy contradiction or indirect reasoning, each beginning with the positing of the opposite of that which one aims to prove. Although not all such proofs are deemed beautiful, the act of beginning with the opposite claim is a highly aesthetic move, in that doing so enacts a kind of feigned indifference or autonomy with respect to the truth of the claim. It is as though the speaker refuses to push hard for the claim, generously offering to indulge the other side ('fine, we'll set my own claim aside, and we'll go along with yours …'), much like Socrates might have done. This discursive move immediately performs a kind of autonomy by setting up a distance between the aim of the proof and the manner or direction of it. This distance makes of the proof an appearance or performance, and we are suddenly invited to perceive the proof as an aesthetic object. Perhaps the conclusion of the proof – that being the negation of the original claim or assumption – wraps the end onto the beginning in such a way that closure and smoothness become *felt* aspects of the proof. The proof becomes a surface folding back on itself – a perfect, opaque ball – then to be cherished and handled like an art object.

However, perhaps these two proofs are considered beautiful because they each produce the unexpected – that being the object that was previously denied existence. In the case of the infinity of primes, yet another larger number is created, literally cobbled together from a collection of commonly held primes, and this number must either be prime or be divisible by a prime number larger than the ones originally posited as the complete, finite set. We hear the speaker say, 'Do you see it now? You said it didn't exist, but I have shown you what it would have to look like.' Another larger prime can always be generated from any finite set of primes; the 'beauty' lies in how the new prime comes forth autonomously through logic, rather than through the labour of its material construction. The proof itself functions in redistributing the sensible and partitioning what is taken to be real. In the second case, a similar creation – that being the irrational number – plays havoc with the ontology of number and emerges, as though by magic, as a new being. The incommensurable – that is, the 'properly' imperceptible – is shown to defy our common sense, and we must recognize that there is a new way of delineating between the sensible (rational) and what was previously taken to be the non-sensible (irrational). We can see that these two proofs deal closely with the partitioning of the senses and the redistributing of what is taken to be 'common' sense.

Netz (2009) makes evident the ways that communities of sense-making in mathematics have been radically different over time, arguing that written mathematics in the time of Archimedes (from about 250 BC to 150 BC) had a distinct style that differed markedly from both that of other ancient Greek periods (including that of Euclid), as well as that of contemporary writing. Before proceeding, seeing as we are going to compare the Alexandrian style with the contemporary one, it is important to point out that Archimedes communicated his mathematics through personal letters and not through journal articles. One may argue that the mode of communication marks the essential difference between the two styles we want to compare. Nowadays, mathematicians are also permitted to communicate through letters (or emails), and their style of writing in these cases differs drastically from that of their more formal writing. However, we think it is still worth comparing the two styles, in part because the Alexandrian letters were the 'common' form of preserving mathematical knowledge, and they communicate complete results through theorems and proofs.

The aesthetic elements that Netz proposes for the Archimedean style are as follows: narrative surprise, mosaic structure, generic experiment and a certain 'carnivalesque' style. Netz borrows the colourful adjective 'carnivalesque' from Bakhtin, for whom it describes a literary mode in which humour and chaos are used to subvert and liberate assumptions associated with a dominant style. These elements are manifest in Netz's reading of Archimedes' account of spiral lines, which is devoted to the proof that the area under the segment of a spiral equals one-third the area of the corresponding circular sector. Although Hardy might have been tempted to judge this proof in terms of its being natural and illuminating, Netz is less concerned with evaluating the result in terms of its aesthetic qualities than he is in analyzing the particular style with which Archimedes relays the result.

Netz argues that narrative surprise can be seen in the very introduction to the problem, which arises abruptly when Archimedes introduces the spiral as 'a special kind of problem, having nothing in common with those mentioned above' (p. 3). Netz shows that this kind of abrupt transition is characteristic of the Archimedean style and reoccurs throughout the letter, even as the proofs are given. It is not until proposition 24 that, as Netz writes, 'the treatise as a whole makes sense', and the enunciation of the result is given 'in economic, crystal-clear terms – the first simple, non-mystifying enunciation we have had for a long while' (p. 10).

Surprise operates here as though we were following an unfolding narrative, and the readers of the narrative are meant to be engaged precisely because they do not know what will happen next. Surprise, in this sense, must, by definition, be completely at odds with a text that unfolds according to a series of deductive implications, because these must 'follow' by necessity. From a pedagogical perspective, promoting surprise may seem counter-productive, given that we surely want the students to understand the ratiocination of the proof rather than submit to the intrigue that compels one to turn the pages of a detective novel. And yet, surprise often accompanies important learning experiences (Movshovitz-Hadar, 1988). Nunokawa (2001), for instance, argues that surprise is a critical factor in good mathematics instruction and that one could plan surprises in lessons by attending to the gaps between conjectures and realizations.

Dreyfus and Eisenberg (1986) claim surprise as one of the important aesthetic qualities of a mathematical problem. Their list includes: 'its level of prerequisite knowledge, its clarity, its simplicity, its length, its conciseness, its structure, its power, its cleverness, and whether it contains elements of surprise' (p. 3). And yet, surprise cannot be scripted or anticipated. All the characteristics of the mathematical aesthetic – clarity, brevity, elegance, conciseness – lack significant impact if a feeling of surprise is not also engendered:

> The conclusion of such a powerful argument tends to contain an element of surprise for anyone who did not know the argument before. This surprise, in turn, is a further contributor to the aesthetic appreciation of the argument; mathematicians, similar to the spectators of a magician, like the unexpected, at least as long as they consider they have a fair chance at understanding the reasons behind the surprising conclusion. The factors contributing to the aesthetic appeal of a solution or proof are thus connected to each other; they almost follow naturally from each other: clarity – simplicity – brevity – conciseness – structure – power – cleverness – surprise. (p. 6)

Stanley (2002), however, has argued that surprise has to be seen as 'an event of emergence' and that those who are surprised must be 'prepared to be surprised', in that surprise occurs only when there is a discrepancy between expectations and experiences (p. 15). The word *surprise* has French and Latin roots in *surprendre* (to over-take) and *prehendere* (to grasp or take, as in *prehensile*), respectively. The word surprise came to refer to 'a feeling caused by something unexpected' in the sixteen century, thus combining the affective with the epistemic. As Stanley suggests, surprise is a deeply relational event, emergent through the interaction of different bodies: '[S]urprises are event-full moments or happenings' (p. 15). In particular, surprises operate through blind spots and the perturbing or subverting of other limitations to the senses.

We find that surprise is thus an event of dissensus, in that the delineation of the sensible – that which is visible, audible, intelligible – is altered and redefined. The logic of consensus is undone when that which was taken to be invisible or inaudible is made visible or audible by an act of dissent that enacts a different kind of sharing of the sensible. Political disruptions of the sensory self-evidence of the 'natural' order of life will always entail an aesthetic component of dissensus, just as art that breaks with the limits of speech and perception will reconfigure the space of political participation. If consensus is an alignment between sense (as sensation) and sense (as meaning), then we use the term dissensus to refer to that which breaks up this alignment. One can see how surprise relates directly to a theory of the body in mathematics and underscores the power of dissensus to motivate the kinds of judgements we have found in the literature on the mathematical aesthetic. We are arguing here that surprise should be dislocated from the individual and seen as an event or happening that recombines heterogeneous materialities and redefines the contours of the sensible. These popular mathematical proofs do not simply allow an individual an expanded capacity to sense – for instance, in being suddenly able to touch something that one could not touch previously. During an experience of surprise, an individual assemblage is literally 'over-taken' by new material assemblages. A surprise is an event through which two or more bodies interpenetrate in new ways, and a new assemblage emerges. Bodies mix and intermingle during surprises in ways that bring forth immanent tensions and new surface effects. Thus, surprise is a crucial facet of the mathematical aesthetic, because it operates through sense (as meaning) and sense (as corporeal activity). But surprise is also a form of dissensus, because it disrupts that which is expected. Surprise is a place or site of breakthrough, making it both what produces new bodies and also what must be presupposed for corporeal activity. Much of the literature on surprise, however, tends to package it as an aesthetic *judgement*, rather than studying the corporeal activity involved in it.

Netz explains that surprise operates in the Archimedean text because there is no obvious thread through the proofs that are offered, and the spiral is not even defined until halfway through the letter. Before getting there, we have a 'surprising sequence going from physics through abstract, general geometrical observations, via the geometry of circles and tangents, and finally, leading on to a *sui generis* study of arithmo-geometry, none of these being relevant to any of the others' (p. 9). Netz sees the extensive use of calculations and physics (Archimedes' spiral requires the motion of two lines for it to be called into being) as a breaking of genre boundaries and the ungoverned sequence of seemingly unrelated material as leading to a style of surprise and mosaic structure that contrasts greatly with the linear, axiomatic presentation found in contemporary mathematics. In addition, in contemporary mathematics, efforts are made to signpost the general structure of the argument, so that the reader knows how different tools – and, especially, different lemmas – are being used. This pedagogical style seems to be completely absent in the Archimedean letter.

Netz claims that Archimedes intentionally chose an obscure and 'jumpy' presentation so as to 'inspire a reader with the shocking delight of discovery, in proposition 24, how things fit together; so as to have them stumble, with a gasp, into the final, very rich results of proposition 27' (p. 14). The Archimedean writing style might thus be described less as being in pursuit of the true or the good, and more as being designed to produce a highly satisfying emotional reaction, much in the same way we expect a good detective novel to work. Furthermore, Netz points to the way in which we can attend to this Archimedean treatise in terms of the novel and somewhat exotic focus on the spiral, which he was the first to study, and which involves boundary crossings not customary in Euclidean geometry, where time and motion are customarily strictly forbidden. By means of these stylistic elements – which are evident in the extensive number of examples of mathematical writing by Archimedes and his contemporaries that Netz provides – a mathematical style emerges that contrasts markedly with the contemporary one. We have already hinted at some of the differences, but it would be misleading to neglect one difference upon which Netz elaborates at length, namely, the way in which Archimedes' mathematical writing style was influenced by, and in turn influenced, the Hellenistic literary style in poetry. It would be difficult to make a similar kind of argument today (unless one wants to consider the works of groups such as Oulipo), but in articulating the central tensions of both literary and mathematical cultures, Netz provides insight into the way in which a different style might have been possible in the past – and, indeed, might be possible again in the future.

In this section, we have explored contemporary mathematical sensibility and the particular forms of consensus that have been articulated by mathematicians. We used Netz's exploration of Archimedean mathematics as a point of comparison that exemplifies a radically different mathematics community of sense – and, in doing so, we have tried to show not only how the current community of sense is a choice (and not a necessary consequence) of the discipline, but we also want to ask what acts of dissensus might look like today. Such acts occur quite regularly in mathematics, in fact, and can be seen perhaps most clearly any time decisions are made about what counts as mathematics or how mathematics is different from other disciplines. One rather highly public example revolved around the Jaffe and Quinn debate (1993) that concerned boundary-making between the disciplines of mathematics and mathematical physics (a debate that would have been of particular interest to Archimedes!). At stake for Jaffe and Quinn was the safeguarding of mathematical rigour against speculation as it occurs in current interactions between physics and mathematics. Their paper provoked much debate, resulting in 16 responses by leading mathematicians in the subsequent issue of the *Bulletin of the American Mathematical Society*. While the issues raised were varied, much of the discussion turned on the separability of conjecture and proof. Common sense clearly distinguishes the two – despite Imre Lakatos's insistence on their fundamental dialectic nature – with proof being the only currency of truth. If conjectures can be published, as well as credited, as they have been by some theoretical physicists, then a new currency will circulate. On a more subtle level, if conjectures can be published, as well as credited, then the contingency and plasticity of formal mathematics will become far too visible.

In the next section, we continue our exploration of the theme of what counts as mathematics and mathematical activity in the context of school mathematics and, in particular, with respect to policy and curricular documents. We will show how the new consensus concerns the politics of becoming in the mathematics classroom.

New standards of curricular consensus

The struggles that have played out in mathematics curriculum reform over the past 20 years are frequently described as being highly political, in that what is at stake are the competing values of the traditional and reform camps. In his analysis of mathematics curriculum, Popkewitz (1998, 2004) focuses not on the politics of math education as it materialized in the so-called math wars, but on the often-neglected politics of pedagogy where the standards of reform are enacted and, as he argues, also produced. For those who have been involved in arguing for alternatives to traditional pedagogies that have disenfranchised large groups of learners and teachers, Popkewitz's analysis may seem counter-productive. Yet, his critical perspective troubles the claims that one practice is better than another, because no pedagogy can be entirely just and equitable. Popkewitz helps us reflect on how the reform movements of the last few decades entail consequences that 'divide, demarcate, and exclude particular children from participation' (p. 1). More specifically, he considers particular reform practices that may actually reduce the 'range of phenomena for scrutiny, action, and critical thought' (p. 18).

Popkewitz organises a significant part of his critique around the term 'alchemy', which he uses to describe the way in which school subjects are formed through a transmutation of academic knowledge, where the governing principle is no longer, say, mathematical knowledge, but now mathematics pedagogy. Brown and McNamara (2011) provide a stunning example of this transmutation in their description of the changing identities of primary school teachers during the years in which the U.K.'s National Numeracy Strategy was being implemented. All research participants were asked the question 'what is mathematics?' in each of their four years of training. Their answers at the outset of their studies were clipped and numerically oriented, but later they broadened to responses such as 'exploring number, exploring shape' and 'comparing multiple solution strategies'. By the end of their training, their statements revealed a conception of mathematics that primarily involved good management of activity and commodified curricular performance standards. The authors make the insightful observation that pedagogical forms (the use of manipulatives or line graphs or drill sheets) came to stand in for the mathematics itself: 'The presentation of the activity seems to provide a way of locating mathematics, yet the activity seems to be clouding the teacher from alternatives. The pedagogical form becomes the mathematics itself such that it is otherwise "impossible to teach"' (p. 113).

Popkewitz (2004) uses the term 'inscription devices' to refer to the kinds of pedagogical forms that Brown and McNamara name in their study. For Popkewitz, alchemy happens through an 'assemblage of inscription devices that translate and order school subjects' (p. 2). The significance of inscription devices is that they make visible the thoughts of a child in ways that make them amenable to governing. As an inscription device that emerged in the 1980s to counteract excessive attention to procedural thinking and memorization of facts and to reflect the nature of the mathematical discipline better, 'problem solving' also changes the features of a child's inner characteristics and capacities that are deemed salient, and it functions to demarcate, preserve and make administrable these features. One can argue – and people have argued – whether or not the particular inscription devices of reform mathematics are 'better', but instead Popkewitz draws attention to the inevitable pedagogizing of mathematics that has ensued. Indeed, in the U.K. context, the alchemy made it easier – in the eyes of the teachers – to teach 'mathematics' (which had become drill sheets), but the opposite seems to hold true in the U.S. context, where reform practices seem to make 'mathematics' (which has now become problem-solving and group discussions) much more difficult to teach. But in both cases, one can trace the way that the mathematics becomes a set of 'commodities exchanged in the educational marketplace' (Brown & McNamara, 2011, p. 126), which are directly linked to particular forms of social regulation.

In the U.S. context, the particular alchemy on which Popkewitz chooses to focus seems especially complex, in that at least some of the educators involved in promoting inscription devices such as problem solving and communication drew their pedagogical forms directly from the philosophy of mathematics. More specifically, it was Lakatos's *Proofs and refutations*, in which an extended instance of historical mathematical practice was compellingly described through an imaginary dialogue theatrically set in a mathematics classroom, that inspired many of the inscription devices. Lakatos criticises the deductivist approach encoded in the formalist philosophy of mathematics, in which mathematics 'is presented as an ever-increasing set of eternal, immutable truths' (p. 142). In contrast, he offers a more fallibilist approach, and he characterizes this approach by describing the methodology of proofs and refutation, which is a general heuristic pattern of mathematical creation that consists of several stages from primitive conjecture to 'proof' to the consideration of counter-examples that result in an improved proof.

The imaginary dialogue presented in the Lakatos book consists of a historically inspired account of the Euler-Descartes formula for polyhedra. The whole method of proofs and refutations centrally involves the creation of putative counter-examples, which have become known as 'monsters' (polyhedra that do not fit the formula relating the number of vertices, edges and faces – V-E+F=2 – such as the cylinder), and the barring of these monsters (which occurs through changes in the definition of terms such as polyhedra). The simple, familiar cylinder now becomes a monster in the face of the desired formula. Does one abandon that beloved relation or make a new partition of the sensible in which the cylinder is no longer a polyhedron? Indeed, the nineteenth-century mathematicians involved in this work decried the twisted, nonsensical monsters proposed by their colleagues, begging for a return to the tamer polyhedra considered (and possibly intended) by Descartes and Euler. Monster-making and monster-barring compromise the autonomy of the polyhedron concept. Finding these monsters, that is, and producing these counter-examples can be seen as acts of dissensus. The sensory disorientation produced by such acts can be deeply disturbing, as Lakatos exemplifies when Delta, one of the characters in his play, recoils in horror from Alpha's 'monster' of nested cubes: 'I turn in disgust from your lamentable "polyhedra", for which Euler's beautiful theorem doesn't hold. I look for order and harmony in mathematics, but you only propagate anarchy and chaos' (p. 21).[5] But once the process of proofs and refutations produces its provisional partitioning, the polyhedron must emerge, autonomous once more, delivering itself back to the real world of tangible objects. Eventually, at least for now, the sensible nature of the three-dimensional objects under consideration (often presented in two-dimensional perspective on paper) succumbs to the imposing austerity of the formula, which itself eventually pursues its own line of flight when applied to topological spaces.

As Pimm, Beisiegel and Meglis (2008) point out, if Lakatos's main argument was that 'progress at the frontiers of mathematics *does not occur* by a deductive process but rather, by the very heuristic process' exemplified in his dialogue (p. 474, *emphasis in original*), there is still a long way to go before anything can be claimed for the *learning* of mathematics. Further, not only is the Socratic-like dialogue not an accurate representation of discourse, it is also not, by any stretch, meant to be an accurate representation of a mathematical classroom. That being said, many mathematics educators have jumped to facile analogies between a long, drawn-out, historical mathematical production involving expert mathematicians producing new ideas and a single classroom lesson involving children learning known ideas. For example, Ernest (1991) argues that the teacher and students should engage in ways identical to those in Lakatos's dialogue, specifically posing and solving problems, articulating and confronting

[5] Lakatos is paraphrasing Hermite, who was writing a letter Stieltjes 'with a shudder of disgust' about the 'plague of functions' that the latter was offering as counter-examples of functions that are continuous but have no derivatives.

assumptions, and participating in genuine discussion (p. 208). Lampert (1990) applies the dialogue to school mathematics in her experiment to test whether the qualities of Lakatos's historical mathematics account could be observed in a classroom setting. The experiment was a success, in that the students 'learned to do mathematics together in a way that is consonant with Lakatos's and Polya's assertions about what doing and knowing mathematics entails' (p. 33). Lampert's work was later used to model and justify – using a Lakatosian dialogue – the NCTM Standards' vision of a mathematics classroom (Yackel & Hanna, 2003).

But the 'doing and knowing' of mathematics was based almost entirely on the *form* of Lakatos's assertions, namely the dialogue between students and teacher. The *content*, which involves the formulating of definitions, the creating of lemmas, the stretching of concepts, the barring of monsters, the bickering between 'students', the questioning of taste, the political and intellectual accusations, and the historical links, were taken to be epiphenomenal. Thus, the policy initiatives took up the book in a way that chose certain forms of participation to be pedagogically valuable. The central disciplinary component of Lakatos's work was left behind in favour of the new inscription device of classroom discussion.

Of course, political actions do not just occur on the level of large-scale policy issues, like 'problem solving'. A much narrower, and perhaps more mundane-looking example, can be found in the introduction of the two-column proof in American high school geometry courses in the early twentieth century. Herbst (2002) shows how this inscription device 'helped stabilize the geometry curriculum by melding together the proofs given by the text and the proofs expected from the teacher' (p. 285). Like 'problem solving', the two-column proof determines the material practices that are deemed salient, while also distinguishing common sense from non-sense and, in doing so, classifies children according to their capacity to perform accordingly. Like 'mathematics for all', the two-column proof aimed to meet the demand that 'every student should be able to do proofs' (p. 285). The two-column proof invokes a very specific material practice, in which arguments are made in rows and columns instead of in a more narrative style. All statements must fit in some cell of the table, and no statement that is unrelated to the properties of that cell can be written. Furthermore, *every* statement has to have a reason. In fact, this format involves a necessary detachment from engaging materially with the diagram, in that the statements supposedly mediate one's interaction with the diagram, because one engages with the notation, or labelling, and the utterances about these labels, rather than the physical markings of the diagram, and in doing so,

PROPOSTITON XIX. THEOREM

106. *If two parallel lines are cut by a transversal, the corresponding angles are equal*

(Converse of Prop. XIV.)

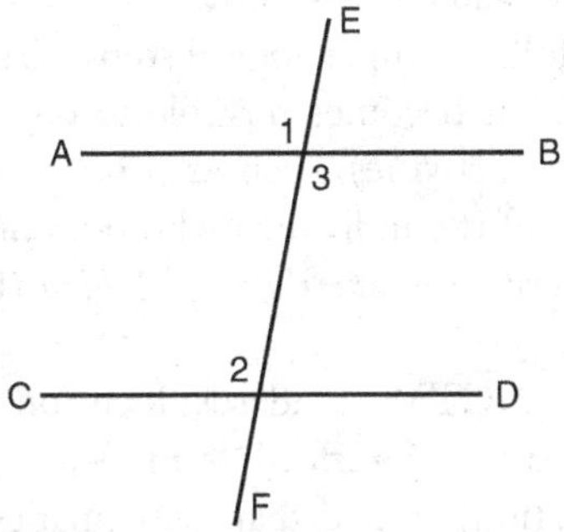

Given parallel lines *AB* and *CD* and the cor. ∡ 1 and 2.

To prove ∠ 1 = ∠ 2.

Proof

STATEMENTS	REASONS
∠ 1 = ∠ 3.	Vertical ∡ are equal..
∠ 2 = ∠ 3.	Alt. int. ∡ of ll lines are equal.
∴ ∠ 1 = ∠ 2	Things equal to the same thing are equal to each other.
Q.E.D.	

Figure 7.1. A two-column proof on corresponding angles. (Photographed by the author from Schulze and Sevenoak, 1913, p. 53.)

the format keeps hands away from the image. It also creates a strict visual divide (the left and right columns) between utterances and justifications, thus performing an extraction that forces the students to divorce reasoning from expression. The impact on the hand and the eyes is immense.

Its material consequences extend to the concept of proof, so that every statement has – and has to have – a reason and vice versa; perhaps most stunningly, the means of discovery appear nowhere and, as such, remain completely separate from the logic of justification. The chain of events is illuminating: Educators decide that it is important for students to write their own 'original' proofs, rather than memorize and copy Euclid's, but then they realize that this is quite difficult, so they find a way of making proof-writing more accessible through the two-column inscription device. The result is that students learn to write two-column proofs in which the *what* and the *why* have been spatially separated. The question Popkewitz invites us to ask is how such alchemy changes the way a child's inner characteristics and capacities are viewed. In the regime of the two-column proof, the divisions between those who can prove and those who cannot is altered radically. If earlier a student could *not* prove because proving was a challenging task, now the blame shifts to the student, who cannot prove because of an inability to follow simple, logical steps. As soon as the problem shifts to the individual, it becomes possible to objectify the learner: 'The mapping of children's activities, such as problem solving, simultaneously creates a mapping of the individual who does not 'fit' or act as a problem solver and is inscribed as *the child left behind* (Popkewitz, 2004, p. 5, *emphasis in original*).

A decade following the NCTM Standards' focus on problem solving, in 2001, *Adding it Up* (Kilpatrick, Swafford, & Findell; a publication from the U.S. Center for Education) argued that mathematics learners should have a 'productive disposition', which 'refers to the tendency to see sense in mathematics, to perceive it as both useful and worthwhile, to believe that steady effort in learning mathematics pays off, and to see oneself as an effective learner and doer of mathematics' (p. 131). While this insistence on seeing sense seems reasonable in the context of well-documented, rule-driven or procedural activity, in which, ironically, 'making sense' is hardly operative, we can also read this as a new inscription device that goes even further and deeper in demarcating, dividing and excluding because of its focus on the individual. It is psychological in the sense that the child's soul is now at stake: '[P]sychological inscriptions focus on the interior dispositions or the *soul* of the child, fabricating the problem-solving child as a particular human kind for pedagogical intervention' (p. 4). Instead of talking about a child's ability to replicate or demonstrate understanding, *Adding it Up* invokes senses, beliefs and identity. There is now a new version of the aesthetic regime at play, in which the alignment between two kinds of senses – making sense, as in being understandable (epistemologically), and having sense, as in being worthy or useful (axiologically) – are what sustains consensus.

As mathematics teachers, we also find ourselves hoping that learners see value in the mathematics they must study in school. We hope this in part because of our own rich and satisfying experiences with mathematics. We also realize, following Bishop (1988), that mathematics carries with it a particular set of cultural values that are most often not made explicit in the mathematics classroom. However, these values are much more complex than 'useful and worthwhile'. Indeed, most of mathematics is *not* useful. Steady effort does not always pay off. Being an effective learner and doer of mathematics might conflict with other ethical commitments. Mathematics is rife with nonsense. Instead of coercing students to embrace the new, more all-encompassing aesthetic regime (an effort unlikely to succeed),[6] why not see what the reigning mathematical aesthetic offers as opportunities for dissensus? Consider the learner who manages to make sense of Pick's theorem while also coming to see it as utterly useless and trivial; or the learner who perceives the usefulness of 'invert and multiply' when dividing fractions but will not make sense of it.

The psychologies of instruction that are invented by reform (be it a two-column proof or problem solving) are designed to normalize the child, so Popkewitz (2004) argues that they are inadequate for the purposes of translating mathematics into curriculum projects. He suggests that translation requires intellectual tools that 'consider the relation between the knowledge (concepts, generalisations) and the cultural practices that enable the production of the knowledge' (p. 27) in such a way as to avoid psychological reductionism. Insofar as Lakatos's *rational reconstruction of history* (his term) focuses on relations that construct the discipline and exemplifies the way in which the discipline grows and changes over time, he can be seen as offering tools for thinking about and ordering the practices of mathematics. One of the important features of mathematical growth to which Lakatos points relates to the way in which 'taste' plays an important role in determining the shape of the proving process. Interestingly, this feature is completely absent in any pedagogical translation of Lakatos's work. And it is arguably one of the crucial points at which the vision of a discussion-based, problem-solving reform classroom breaks down: How does one decide, in the absence of any purely logical means, which definition will be chosen, which claims will be embedded into lemmas, which mathematical monsters will be explained away?

As Popkewitz and many other scholars have persuasively argued, schooling is a form of cultural politics, whereby particular social agendas become entrenched as common sense. School mathematics – as a 'high-status discipline' (Ahlquist, 2001, p. 27) – plays a significant role in the production and validation of what is taken to be common sense. The extent to which one masters the alphanumeric practices associated with this common sense correlates in large part with the acquisition of cultural capital. This

[6] While research has shown correlations between belief in the worthiness and usefulness of mathematics and strong achievement, as well as between self-efficacy and strong achievement, it is extremely misguided to assume that there is a causal relationship or that is it known how such a 'productive disposition' might be taught or instilled in the mathematics classroom.

cultural capital is distributed according to socio-economic class, 'race'/ethnicity, gender, (dis)ability and other social categories (Gates, 2002). As a 'critical filter' implicated in the social stratification of communities, school mathematics becomes pivotal in the social structuring of students' lives (Moses & Cobb, 2001).

Skovsmose (1994) uses the term 'critical mathematics education' to describe attempts to address this fact and to reconceive school mathematics as a site of political power and ethical contestation. Various proponents of critical mathematics education have pursued this agenda in different ways.[7] Skovsmose and Borba (2004) are careful to suggest that the critical approach must always attend to the 'what if not' of school mathematics – that we must investigate the possible, consider the otherwise and explore 'what could be' (p. 211). This approach 'confronts what is the case with what is not the case but what could become the case' (Skovsmose & Borba, 2004, p. 214). Similarly, Pimm (1993) encourages a shift 'from should to could' in teacher education, arguing that the lust for change (which focuses on how teachers should teach and students should learn) ignores the sense of the personal and the possible. Given that schooling is a form of cultural politics, we need to study mathematics sense-making in terms of a 'distribution of the sensible'. Our inclusive materialism is an attempt to do so.

In this chapter, we have pointed to several different acts of dissensus, beginning with that of Sean, who tried to shift the binary distinction of even and odd numbers into a tripartite one, in which numbers could be even, odd, or both even and odd. We also elaborated on how surprise and counter-examples entailed dissensus by significantly shifting our ways of sensing, thus calling into question what makes sense in mathematics. Even if Sean's classmates and teacher had joined his revolution, a new regime would have inevitably settled in (albeit an interesting one, where, say, rectangles might be both squares and parallelograms or quadratic functions might be both first and second degree). As consensus grows, every shocking new counter-example (the sphere as a polyhedron!; the everywhere continuous but nowhere differentiable function!) gets tamed into the 'obvious' and the 'clear'. When Brown and Walter (1983) advocated for problem

[7] For example, by generating a socio-political ethics of mathematics education (Skovsmose & Valero, 2002; Valero, 2004; Valero & Zevenbergen, 2004); designing new mathematics curricula that address social justice issues (Mukhopadhyay & Greer, 2001; Gutstein, 2006; Tate, 2005); examining mathematics teacher identity and resistance to social justice pedagogy (de Freitas, 2008a de Freitas, 2008b; Rodriguez & Kitchen, 2005; Walshaw, 2004a Walshaw 2004b; Zevenbergen, 2003); and deconstructing the linguistic strategies unique to school mathematics that inhibit increased participation (Adler, 2001; de Freitas & Zolkower, 2009; Morgan, 2006).

posing in the mathematics classroom, they offered the radical, and surprising, idea that mathematics education could also involve posing problems, not just solving them. Allowing students to pose problems, just like mathematicians do, changed the nature of who was in control of the questions in the classroom. But it turned out to be rather hard to manage at scale. So, in the first step towards consensus, the 'what-if-not?' means of posing problems was offered to teachers and students, but it was all too often realized, unfortunately, as a set of prescribed alterations to a given problem situation. Then, most ironically, researchers began to study whether problem posing could work in the service of other (more desirable) goals, so that the very idea of posing problems folded into an existing consensus around school mathematics.

Dissensus eventually produces a new consensus. The question thus becomes: How might dissensus-producing ideas be kept lithe and fleeting, so that they escape becoming part of the common sense while remaining meaningful for a community of practice? We have offered surprise (which is often produced by counter-examples) as a sign that there is opportunity for much-needed dissensus, one that has a progeny within the discipline of mathematics. Surprise is often short-lived. It often occurs at the local level and is produced by disturbing expectations, be they mathematical or pedagogical: for example, by asserting that a line is a circle with the centre far, far away; by imagining that odd numbers do not exist; by teaching quadratic functions before linear ones or tangent before sine and cosine; by letting parallel lines meet; by asking for a wrong answer with an explanation; by showing how judgements of simplicity are driven by context; etc. The aim is to perturb, if even only temporarily, what is taken to be common sense and who is assumed to possess it.

8 The virtuality of mathematical concepts

Our aim in this chapter is to show how a theory of embodiment must simultaneously rethink the body *in* and *of* mathematics. In other words, we want to show how mathematical concepts function as material agents in mathematical activity. Trying to understand how seemingly abstract concepts become parts of body-assemblages does not simply involve locating the concrete sensorimotor activities that supposedly give rise to abstract mathematical concepts. When Lakoff and Núñez (2000) describe the container metaphor from which the mathematical idea of 'set' emerges, they remain committed to an immaterial mathematical concept of set. Moreover, as we will argue, treating all concepts as metaphorical in relation to the 'real' reinforces the divide between the mathematically abstract and the physically concrete.

In this chapter, we return to Châtelet's notion of the virtual. We focus on how the virtual attends to this mobile, generative feature of mathematical concepts, allowing us to study the potential energy of a concept in terms of the kind of work it can do. Our aim is to lay the groundwork for a new way of thinking about concepts within the mathematics curriculum.[1] We spend considerable time describing and exemplifying what Châtelet means by the term *virtual* in the context of mathematics, both because it differs in important ways from everyday uses of the word, and because the term might be seen by some readers as entailing a kind of neo-Platonism, from which Châtelet

[1] Vygotsky (1934/1986) is the only psychologist to have argued that all concepts, even mathematical ones, begin as abstract and move towards the concrete – an antipodal view to the much more commonplace assumptions of Piaget. Although we are partial to Vygotsky's work, his inversion of the concrete to abstract development does not, in our opinion, go far enough in treating the special nature of mathematical concepts and their awkward material virtuality.

dearly wants to distance himself.[2] In the second part of the chapter, we explore how school mathematics concepts might be virtualized in the spirit of Châtelet.

Bridging the mathematical and physical worlds

Châtelet's interest in the virtual originates in his dissatisfaction with both Plato's and Aristotle's solutions to the paradox that emerges when the mathematical and the physical are considered to be ontologically distinct. The paradox concerns the impossibility of a *causal* relationship between the mathematical (as that which pertains to eternal truths) and the physical (as that which pertains to the contingent and the real). Plato's solution was to posit a timeless being that 'recollects' the ideal forms of mathematics. Aristotle's solution was to infer the existence of some 'final cause' that can ensure cohesion between the two natures.[3] Much later, Galileo courageously proposed getting rid of Aristotle's teleological explanation, instead suggesting a link between the *genesis* of mathematical and physical concepts. What could this link be? Châtelet pursues this question with an alternative solution, claiming that the *virtual* – a state of being that is *both* physical and mathematical – provides the necessary link between the two realms; in doing so, he challenges the Aristotelian embargo against motion in mathematics.

Châtelet conceives of the virtual as the necessary link that binds the mathematical and the physical together in mutual entailment: In this he follows Deleuze (1994),[4] who asserts that '[t]he virtual must be defined as strictly a part of the real object – as though the object had one part of itself in the virtual into which it is plunged as though into an objective dimension' (p. 209). In other words, mathematics cannot be divorced from 'sensible matter', and it is the virtual dimension of this matter that animates

[2] The term is also used to refer to the 'virtual reality' offered by digital technologies, in which reality is *potentially* present but not actually so – that which *seems* real but is not. Working within media studies, Burbules (2006) describes human experience of the virtual in terms of the 'feeling of immersion', which involves an extension or elaboration of that which is present in experience. In this sense, it *feels* real but is not. These notions of the virtual continue to maintain a distinction between the real and the non-real or at least a boundary between material and immaterial worlds.

[3] As Châtelet writes, Aristotle later sketched out another solution to the problem, which Leibniz studied and eventually developed along the lines described in this chapter.

[4] We use 'follow' here because Châtelet is known to have read Deleuze and enthusiastically followed some of his interests in the link between philosophy and mathematics; Châtelet also cites Deleuze in his 1993/2000 book. Their names appear together in a book devoted to their joint (along with Guattari) interest in diagrams (see Dupuis, 2012).

the mathematical concept. Mathematical entities are thus material objects with *virtual* and *actual* dimensions. Mathematicians engage these entities in thought experiments, but these are not the disembodied mental ruminations with which we typically associate mathematical thinking; rather, they are physical experiments in which objects are set in motion, cut out from the plane, plunged into singularities, stretched around points of inflection, packed into Abelian groups and so on. As Châtelet (1993/2000) suggests, '[o]ne could even say that the radical thought experiment is an experiment where Nature and the Understanding switch places' (p. 12). When geometric points are seen as mobile material bodies, the mathematician can direct them to new places and even lift them off the Cartesian plane so that they engender new kinds of entities in the complex field. In other words, within Châtelet's materialist account, mathematical concepts engage in a process of becoming, a process that binds them to the actions of mathematicians.

Châtelet credits Leibniz with properly recognizing the dynamics of virtuality. Leibniz was critical of the Cartesians of his day who conceived of geometric points and curves as occupying a rigid, grid-like space. Leibniz's response was to propose a more vitalist, or muscular, conception of matter, whereby space and action are merged through a 'generalized elasticity' that functions to 'fluidify space' (p. 25). For Leibniz, space is conceived as a flexible, folding and animated substance, in which all positioning is relative, and any sense of distance or measure is also relativized. In such a world, points are no longer positions, but rather 'explosive forces' that bring forth new entities (like circles, lines, intersections). Working with points as material forces of motion troubles the fixity and rigidity of the point or the line and, perhaps more importantly, enables us to imagine points as powerful devices that extend beyond their apparent givenness. As Châtelet shows, Leibniz conceives of the geometric point as potential movement, not something that is merely designated (and labelled) as belonging to a space.

Although we do not wish to enlist all of the tenets of the Leibnizian metaphysics, we do want to rethink the coupling of matter and mathematics, and Leibniz's relational space allows us to do that. This kind of approach to space resonates with the materialism we discussed in Chapter 2 by binding the geometric point to matter in ways that break with Cartesian dualism. It also breaks with Newtonian conceptions of space, because the geometric point is not conceived within an absolute frame of reference, as it would be for Newton, but is encountered only relationally within relative measures of space. This point partakes of the material world: This is a mobile point, carving up matter and colliding with other points. And it is precisely this mobility *within* the point – this mobility of the abstract concept – that allows us to begin to grasp how the concept partakes in the virtual dimensions of the material world.

Instead of seeing points as things to be designated – since, according to Châtelet, 'designation assassinates the virtual' – Leibniz considers points as forces of motion. Working with points as forces of motion enables a kind of 'experimental provocation' that, according to Châtelet, leads to encounters with the virtual (p. 15). Consider the simple example of a point on a line. For Châtelet, there is not much interest in contemplating a point $x = 1$ as a staid and detached object confined to a rigid line. Such a point is fixed, with nowhere to go, nothing to do. But when the point is animated within a more 'muscular' conception of space, perhaps becoming a pole or singularity (of, say, the function $y = 1/(1-x)$), the point becomes alive, thus regaining its virtuality. The point is nothing in and of itself; it becomes something through its creative potentiality, as Châtelet evokes in this fervent account: 'construire des mathématiques, c'est construire, en quelque sorte, une manière de faire "fleurir les points"; de différentes façons, on peut avoir une botanique de topologie algébrique ou une botanique d'analyse fonctionelle' (constructing mathematics is constructing, in a way, a manner of making points flower; in different ways, we can have a botany of algebraic topology or a botany of functional analysis, p. 11, *our translation*). This *becoming* of the point is also evident in Cauchy's residue theorem, which enables the evaluation of a path integral of analytic functions around closed curves that contain singularities by wrapping up those singularities in a tight loop and pulling them off. Instead of being a fixed point on the surface, the residue becomes a force of resistance, preventing the paths from closing in too tightly on themselves.

Châtelet admits that the virtual is difficult to explain and difficult to understand, but it is nevertheless a powerful device for thinking about mathematics. In discussing the virtual, he leverages two pairs: the virtual/actual and the possible/real. He uses these two pairs to describe the complex processes of becoming in mathematics. In fact, he envisions every process of becoming mathematics as entailing two aspects, or movements, one between the virtual and the actual, and the other between the possible and the real. Mathematical activity thus involves both *actualizing the virtual* and *realizing the possible*. Together actualization and realization bring forth something new into the situation, but realization plays by the rules of logic (inductive or deductive inferences), while actualization involves a different kind of determination, one that generates something ontologically new, like the complex number or the point at infinity of projective geometry – neither of which emerges from the rules of logic alone. It would seem that many mathematical objects can be thought of in both ways, as logically and materially determined. The circle as the set of points equidistant from a given centre 'realizes a possible', namely, the structural constraint by which classical logic determines the distribution of the points. But in addition, there is an actualizing of the virtual because the circle is *also* the trace of a point constantly being pulled to a centre by an equal reciprocal force as it moves forward.

These two pairs of processes occur throughout mathematics: The logic of the possible aligns with rules of inference, while the onto-logic of the potential is what brings forth the new. And yes, these two processes are at times in tension, but the history of mathematics offers ample evidence that such tension – often in the shape of a paradox – has been incredibly fruitful. Our emphasis here is on the way we might attend more carefully to the virtual and the process of actualizing, given that realizing the possible is a much more common understanding of mathematics. When we pay attention to the mobility of the point, for instance, we also tend to get better at imagining the potentiality of the now more-animated concept of point. In other words, we tend to animate the concept and continue to imagine new configurations, alterations and mutations. For instance, ease up on the force of the point that was tracing a circle little by little, and you have a spiral gesture. In operating alongside any logical determination, actualizing the virtual involves 'an intrinsic genesis, not an extrinsic conditioning' (Deleuze, 1994, p. 154). As an intrinsic genesis, the virtual is generative of mathematical forms but not determinate of such forms.

It is essential to resist the tendency to imagine the virtual as a *form* that is somehow buried in matter and waiting to be conjured or evoked. If the virtual is simply the articulating or actualizing of an *a priori* invisible form, then we must head back to Platonism and the premise that ideal mathematical entities find their inaccurate, fuzzy realizations in the real world. The virtual is not a realm of forms to be mirrored in the physical world. *The virtual does not resemble the actual.* This lack of resemblance is crucial in appreciating the power of the virtual to disrupt the common Platonist epistemology about mathematics. Moreover, because the virtual partakes of the physical, the paradox of causality that confounded Plato and Aristotle is circumvented. Instead of the actual being a copy or tracing of the virtual, '[t]he actualization of the virtual is instead a genuine creation of something that corresponds to singularities and differential relations but does not resemble the virtual' (Cheah, 2010, p. 86). Thus, novelty, genesis and creativity (rather than conditions of possibility) are fundamental concepts in a theory of virtuality.

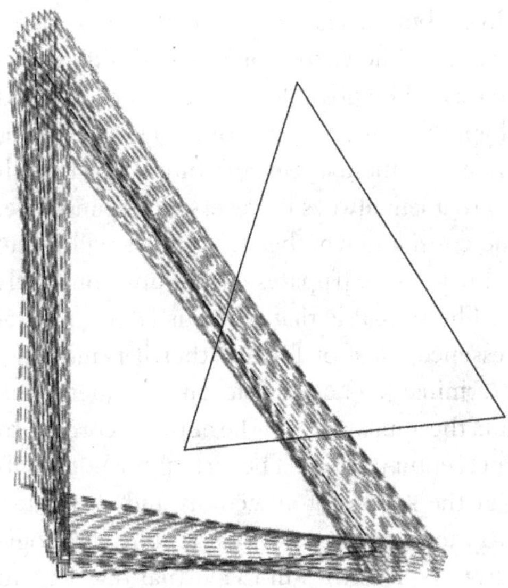

Figure 8.1. The rigid, inert triangle cut-out from the multiplicity of indeterminate pre-figures.

Attending to processes of actualization demands that we reconceive entities such as the triangle or circle less as static figures and more in terms of their virtual motions. In other words, the virtuality of a diagram consists of all the gestures and future alterations that are in some fashion 'contained' in it. A triangle, for instance, does not exist as a rigid figure or as a sign perched in space, but rather it exists as a mobility or set of gestures. One might think of a geometric figure like the triangle as a provisional cutting-out, or articulation, of the indeterminate multiplicity of pre-figures (see Figure 8.1). It may be helpful to think of the virtual triangle that takes on a multitude of locations, sizes and shapes that cannot be fixed in space or time, only one of which ever gets cut out, at any given time, into a triangle diagram. In this way of thinking, the rigid, inert, paper-based triangle as we know it – along with its necessary and sufficient definition – emerges from a boundary-making practice that stops motion. At the same time, the particular articulation of triangle will have a host of materializing effects: it might make the moving triangle seem odd or uncanny; or require certain physical tools to recreate it (drawing implements and flat surfaces for inscribing); or make the figure seem like it is independent of human action or impossible to stretch and transform into a monstrous configuration.

Both the real and the possible change over time (what we thought was real before turns out not to be – whether in love, politics, etc.; what was impossible becomes possible), but processes of actualization – like creation – have an irreversible nature. The virtual, once actualized, reconfigures the world in irreversible ways. The possible, on the other hand, has an 'abstract' side to it, being a hypothetical or subjunctive way of engaging with the world (as in the opening statements, 'Given a function $f(x)$ such that...' or 'Let G be a group...') that can always be reversed. For instance, possibility is involved when one considers whether two circles will intersect, while virtuality is involved when one palpates circles until one feels the joints where they intersect. The virtual is that which is latent in matter – but what is latent is not essence, ideal or form; rather it is mobility, vibration, potentiality and indeterminacy. The possible, on the other hand, is that which structures and limits the appearance of the entity according to current rules of inference and perceptual habits. The virtual (or potential) pertains to the indeterminacy at the source of all actions and all matter, whereas the possible pertains to the compliance of our actions with logical and related perceptual constraints. It is thus in the virtual that we can see mathematics as part of a Baradian indeterminate assemblage of potential points and actions of cutting, excavating and folding. When Châtelet speaks of the physico-mathematician, he speaks exactly of this assemblage of matter and meaning.

Many scholars have thought of Archimedes as one of the earliest and most accomplished physico-mathematicians (see Netz, 2009). This reputation is based in part on Archimedes' non-conformist and unabashed use of tools, such as the sliding ruler that he used for *neusis* constructions, which was not part of the sparse 'tool' set of straightedge and compass that was seemingly admitted by Euclidean geometers. It is also based on the particular kind of mathematics involved in some of Archimedes' great results, such as the quadrature of the parabola, in which he finds the area of a parabolic section by balancing each tiny slice of a triangle with a tiny slice of the parabolic segment around a virtual fulcrum. We discussed Archimedes' proof style in more detail in Chapter 7, but here we want to emphasize that the method of exhaustion that he used in the quadrature of the parabola entails a material approach to the mathematical entities with which he was working. For Châtelet, this is not really a matter of Archimedes treating the triangle and the parabolic segments *as if* they had weights that could be balanced using a fulcrum. The virtuality that is so central to Châtelet demands more than analogies or metaphoric relations to the real. In Netz, Noel, Wilson and Tchernetska's (2011) study of the most recent Archimedian palimpsest, there is good reason to believe that the kind of mobility we have been discussing was indeed at the heart of Archimedes' work, in that those lines he was comparing – those tiny slices – had to be set in motion to sweep out the surface of each shape, for only through such motion or sweeping gestures could Archimedes operate on the continuum and *actual* infinity, which, according to Netz, was precisely what he was doing.

Focus on the motion of mathematical entities has been controversial over the years. When dynamic geometry software was first introduced in the early 1990s, several mathematicians publicly criticised it for implying that the same point can be both here and there on the screen. The triangle, they claimed, is exactly its static representation on the page and not a moveable, changeable shape. Their criticism is part of a long-standing history of the detemporalization of mathematics. Plato had Socrates attempt to undermine any suggestion that motion, time and action are elements in geometry, because these would be indicators of an unwelcome human presence.

> Their [geometers'] language is most ludicrous, though they cannot help it, for they speak as if they were doing something and as if all their words were directed towards action. For all their talk is of squaring and adding and applying and the like, whereas in fact the real object of the entire study is pure knowledge. (Plato, in Molland, 1991, p. 182)

Russell (1903) protests the description of objects in motion for lending them unwanted material agency:

> To speak of motion implies that our triangles are not spatial, but material. For a point of space is a position, and can no more change its position than a leopard can change its spots. The motion of a point of space is a phantom directly contrary to the law of identity: it is the supposition that a given point can be now one point and now another. (p. 405)

One can hear in this quotation a commitment to an ontological binary where geometric points exist in some immaterial geometric space beyond the material world. For Châtelet, however, the point is always becoming a point, always caught up in a process of actualizing. If we are able to sustain its mobility and potentiality, the point could carve up matter in new ways. For Leibniz, who was working on differential calculus, the 'characteristic triangle' (see Figure 8.2) that he drew on the page was a triangle animated with potential energy, capable of shrinking or shifting by infinitely small quantities. The triangle sustained this virtuality by operating as the motive

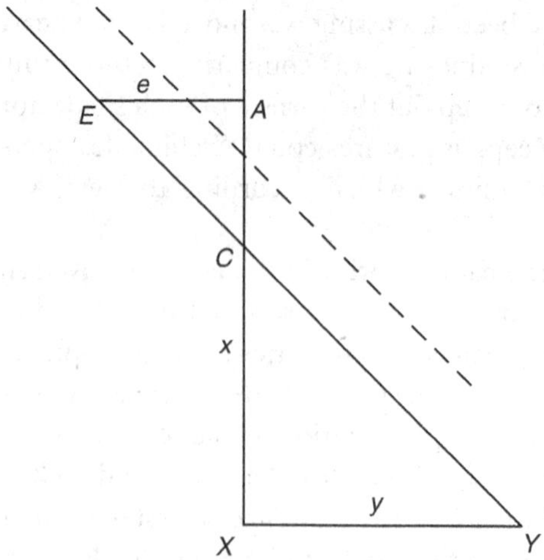

Figure 8.2. Leibniz's characteristic triangle.

engine of differentiation. The triangle was – and had to be – a trembling ambiguity. The extended segments that form the triangle are allowed to vanish, and yet the differential relation between the triangles persists through this movement. Just as with Barad's emphasis on the relational ontology of quantum physics, the differential is a mathematics of relation rather than related. In the following diagram, which is based on a similar one by Leibniz, two similar triangles – ACE and XYC – are put into varying relation through the ratios of their side lengths, operating through the moving dotted line, as it goes from C to A.

Our inherited mathematical ontology can make this kind of thinking difficult. Indeed, Châtelet (1993/2000) complains that we often 'massacre' the virtual by forcing it back into the real or the possible. For example, a frequent move in a calculus lesson is to invite students to think of *dx as if* it were a small increase and to think of the characteristic *dx*, *dy* triangle *as if* it could become a little smaller. Instead, Châtelet implores us to say 'le triangle n'existe qu'en tant qu'il y a des triangles virtuels autour de lui' (the triangle only exists insofar as there are virtual triangles around it, p. 9, *our translation*). Perhaps the dynamic triangle on the computer screen assists us in the sense that one can say that the digital triangle exists only insofar as it can become any or all triangles (which one can achieve, on the screen, by continuously dragging its vertices). When teaching projective geometry, we similarly venture into the apparent 'as if', inviting students to consider the point at infinity *as if* two parallel lines could eventually meet. When Coxeter (1974) speaks of the point at infinity, though, he does not timidly point at the possibility of such a thing but grasps it and stretches it into various configurations. He forges that point exactly at the spot where two parallel lines 'never fail to meet' (p. v). We can hear in his words an embodied encounter with the virtual. And yet, like we all do in mathematics, he ultimately submits the ontological to the logical, foreclosing on the virtual, retreating to the metaphoric and suggesting that we treat these lines *as if* we are looking along a straight railroad.

Châtelet also sees virtuality at work in Abel's famous theorem concerning the indefinite integration of a rational function R(x,y) (where x and y are related by a polynomial equation, $f(x,y) = 0$, which defines an algebraic plane curve, C). For some cases, such an integral is easy to calculate, and there is no need for Abel's theorem, but for elliptic integrals that cannot be solved in terms of elementary functions, a new strategy is needed. The integral of R given by f produces a line integral on the Riemann surface defined by f. (Abel would not have used the term Riemann surface, which refers to a kind of deformed version of the complex plane.) The Riemann surface of a nice integral will have the topology of a sphere, but elliptical integrals might have the topology of, say, a torus, in which case all paths of integration are not connected. Abel provides a way of solving such integrals by focusing on the points of discontinuity of the line integral and imagining the alternate paths passing through them, as in Figure 8.3 where the dashed lines describe different paths – which *can* be integrated – between points of discontinuity.

Modern formulations of this theorem speak of the curve as being 'given' – much like a point might be given on the plane, like an amorphous, abstract thing – and of new paths being found between unwanted points of discontinuity, so that they can be integrated. This, according to Châtelet, masks Abel's sense of the curve as a modulating force or field of

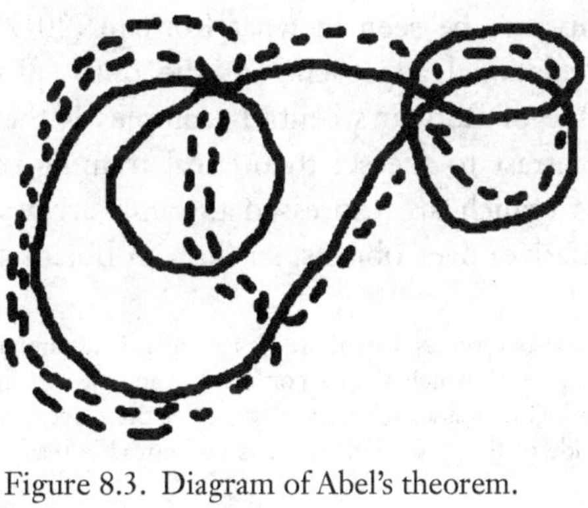

Figure 8.3. Diagram of Abel's theorem.

potentiality with contracted singularities that will become instrumental in defining the integral. The points of discontinuity are forces, not markers, that propel the curve into new, integrable journeys. Indeed, Châtelet wants to see these curves as having '*des démangeaisons*' (itches) at the singularities that are scratched by mathematicians in order to follow new paths of integration. These itches are thus where new objects – which are never only physical, nor solely mathematical – are born. When he writes in terms of 'itches', 'cuts' and 'excavations', Châtelet is not simply being metaphorical. These terms, evocative as they are, are meant to counter the more common static and immaterial discourse of mathematics, where things are given – untouchable but already existing in some determinate form. They are meant to underscore the material nature of the concept.

By focussing on concepts that have evolved out of problems, Châtelet leaves out – in his examples – the mathematical work that is more 'theory building' in nature, such as category theory or the Langlands program for relating algebra and analysis. Should we conclude that Châtelet's notion of the virtual does not apply in such work? It depends on how the theory-building is done. He criticises the set-theoretic foundations of mathematics, deriding its obsession with identity, as does Rotman (2012), who espouses a material semiotic approach to mathematics. The modern set-theoretic approach is championed by the Bourbaki group, who rewrite mathematics in such a way that 'nothing – no notation, definition, construction, conjecture, concept, theorem or proof – is allowed to refer to or invoke or rely on any attribute, body or process of the physical world, not least any reference to the mathematician's corporeality' (Rotman, 2012, p. 249). The contrast between Châtelet and the Bourbaki could hardly be sharper; not only are Bourbakian objects primary and relations secondary, but objects are self-contained, isolated entities, knowable without reference to any other object and without reliance on pictures or diagrams.[5]

A more Châteletian sensibility can be seen in what Rotman (2012) identifies as post-set-theoretic 'relational approaches to the onto-epistemology of objects' (p. 250). One of Rotman's central examples is that of category theory, which, in contrast to the set-theoretical framing of mathematics, privileges relations (which are expressed through arrows, which are composable and associative) over objects, so that 'an object is

[5] The modern set-theoretic approach also reverses Klein's 1872 Erlanger Programme of attending to relations (the symmetries to which shapes conform) over objects (the shapes). Moreover, it mocks Poincaré's (1905) assertion that only the relations between things are knowable and that '[o]utside of these relations there is no knowable reality' (p. 2).

understood relationally, through external difference, not as an autonomous, internally structured entity; it is known and constituted entirely in terms of the arrows entering and exiting it from other objects' (p. 254). A category is a collection of objects (in this case, sets) together with arrows from one object to another (in this case, functions on the elements of the sets). Interestingly, as Rotman points out, objects of categories can also be categories themselves. This avoids the set-theoretical obsession with identity and effectively enables any new object to be defined relationally, as indeterminate multiplicities, as part of a 'mathematical universe of things-in-formation' (p. 255). Rotman concludes that in category theory, a mathematical object is thought 'from the outside, in a "bio-social" register from species to individuals, and not as an isolated, self-contained entity whose relations to others proceed from the inside out' (p. 254). Given that the arrows of category theory essentially and by definition involve operation, power (*puissance*) and movement, there is an inherent virtuality to categories. Furthermore, given that for any mathematical structure, there is a corresponding category (Goguen, 1991), finding the corresponding structure amounts to joining with that virtuality.

Châtelet is not just a historian of mathematics, but also a passionate critic. This is evident not only in his pleas for the virtual, but also in his ruminations on mathematical notation. He describes, for example, the arrows of category theory as the '*revenge* of the hand' (Châtelet, 1993/2000, p. 36, *emphasis in original*), which can be found in this and other mathematical notation practices. The arrow can be seen, like the diagram, as capturing the gesture mid-flight, where the hand expresses the mathematical relationship at play. The notation used in knot theory can also be seen as exerting the hand's revenge.

Châtelet is interested in the co-penetration of the notation (as image/diagram) and the calculation (as action). For example, in the notation used to sum the first 100 natural numbers $\left(\sum_{1}^{100} n\right)$, the action involves calculating $1 + 2 + 3 + \ldots + n$, and there is no co-penetration, because the set of indices 'remains neutral' while the indexation 'remains completely external to the development of these calculations'. In other words, this notation is 'captive to a linear successiveness, x_1 then x_2 then x_3', in which what you say is exactly what you do (2006, p. 35). In matrix algebra, the notation $\sum M_l^i M_j^l$ is used to announce the calculations involved in multiplying two matrices (first column, first row of matrix A times first column, first row of matrix B plus first column, first row of matrix A time first column,

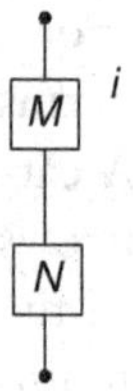

Figure 8.4. Notational rupture for matrix multiplication.

second row of matrix B, and so on). Châtelet compares this notion, which also remains captive to a linear successiveness, with the more 'compact' reading of the notation in Figure 8.4, where the sum is no longer a 'drone', but a connection of the two matrices. This notation refuses to isolate the calculation from the multiplicative allusion of the diagram.

In the end, Châtelet wants to see the notation woven up with the action, not just recounting what must be done, but also suggesting new ways of doing. At its best, notation should function in a delicate equilibrium of icon, index and symbol (using Peirce's categories) or, mathematically speaking, as 'image, allusion and calculation' (p. 41). Compare the notation used for addition in the Egyptian *Ahmes* papyrus (a pair of legs walking forward) with the late fifteenth-century use of 'p' by Chuquet and Pacioli (for plus or più) and the modern use of + (first used by Oresme in the early fifteenth century). The modern notation is less allusive than the Egyptian one, which suggests motion along a path, but more so than 'p', which focuses on calculation. All succumb to sequentiality by insisting on a particular order for addition rather than pointing to its symmetry – and perhaps to a new way of doing addition. (In the *TouchCounts* application described in Chapter 6, addition is done by pinching together groups of numbers – a gesture that avoids the temporality of the notation examples described in this chapter and that might give rise to new notational possibilities in a less alphanumeric age of mathematical communication.)

Might it be worth working directly on our notation in the classroom, much like we work on definitions? Might working on notation (attending to it, reformulating it) involve mathematical thinking? From Châtelet's point of view, certainly! Notation can disrupt the linear flow of the written word – the sequentiality and passivity of mathematical text – but in doing so, it modulates that text, producing new assemblages of meanings. As Rotman (2000) suggests, mathematical inscriptions are not only, or even primarily, notational systems, but rather operate through 'metonymy and synecdoche' (p. 42). They mutate material relations and infuse notation with force and affect. They are a form of kinematic writing, capturing potential movement and affective forces outside of speech. They operate through a refusal to be coded in speech, always supplementary to the system of verbal reference, and yet powerfully effective as highly situated, embodied performances.

For Châtelet, notation and diagram are not so distinct, each potentially playing a significant role in mathematical ontogenesis. While our experiment in Chapter 3 revealed the potential of diagrams in classroom-based mathematical activity, we have paid far less attention to classroom-based notational practices. Notation abounds in school mathematics, but it is often used and seen in passive, indexical ways. One counter-example can be found in Hewitt's (1999) approach to working with algebraic notation, in which, for example, parentheses are activated in a blackboard dance of chalk taps, voice modulations and gestures. We see this kind of focus on and virtualizing of notation as a fruitful area for future research.

We now explore the notion of virtuality of mathematical concepts in the context of school mathematics, in terms of both the sequencing of concepts in the curriculum and the nature of conceptual development employed in textbooks. Inasmuch as Châtelet wants to virtualize mathematics concepts, we want to explore how the same can be done for school mathematics concepts and, if so, what consequences doing so might have on their selection and ordering in the curriculum.

Curriculum mapping of concepts

Using Châtelet's notion of the virtual, we have found a way to define an ontology of mathematics that is consistent with our inclusive materialist perspective. Through the virtual, mathematics partakes of the material world; indeed, mathematics and matter are mutually entailed. The mathematical body comes into being through actualizing the virtual – through gestures, diagrams and digital networks, *we become mathematics*; we incorporate and are incorporated by mathematics. The assemblage grows. Now, we might ask, what does that mean for mathematics learning? Can we find the latent virtualities in the mathematics curriculum and also reanimate the bodies of ossified mathematical concepts? This will be challenging if the virtual is provoked only rarely in major turning points of the discipline. If, however, the virtual is woven through the fabric of mathematics, as Châtelet would argue, how might we recover it, along with the material concepts that have been robbed of their bodies? In this section, we try to show how one might re-examine familiar mathematics concepts in terms of their latent virtualities.

Within mathematics education, concepts are currently the currency of the curriculum, which seeks to specify which concepts should be taught and in what order. Sometimes the curriculum is also described as being ordered in terms of topics or strands and even, in higher-level mathematics, in terms of theorems or theories. In mathematics education research, the word 'concept' has been associated with the notion of a schema (Skemp, 1979), which is described as a 'mental entity' (p. 114), as well as a 'network of connected concepts' (Olive & Steffe, 2002, p. 109). Etymologically, concepts are things that can be taken in (the mind). Vygotsky (1934/1986) describes concepts more functionally as 'taking shape in the course of a complex operation aimed at the solution of some problem' and insists that a concept 'is not an isolated, ossified, and changeless formation, but an active part of the intellectual process, constantly engaged in service of communication, understanding and problem solving' (p. 98).

We can also learn about the nature of concepts by examining how they are deployed through various metaphors of the curriculum. At a basic level, the scope and sequence metaphor for curriculum involves an ordered list of concepts, which is often taken to be the 'natural' order that should be followed when moving on to higher-level concepts without leaving any holes or gaps. For example, in North American curricula, the 'natural' order dictates that addition should come before subtraction, which serves to differentiate these two operations as two distinct concepts (albeit related). However, in China, addition and subtraction are taught together, as a unity of three numbers a, b and c, in which $a + b = c$ *and* $c - b = a$. This might suggest that addition and subtraction are part of the same concept. The curriculum might therefore define concepts in terms of particular and distinct ideas that can build on each other, independent of the kind of learner involved.

Beyond the basic scope and sequence metaphor, Dietiker (2012) writes that the mathematics education literature offers three additional metaphors of curriculum: *curriculum map*, *learning trajectory* and *story*. The first was used by Dewey (1902) to distinguish between the 'logical' organisation of content and the everyday, 'psychological' experiences of the learner. This metaphor encompasses more than scope and sequence in the sense that it invites reflection on the way in which the parts of the terrain relate to each other, including the particular students involved. Like scope and sequence, it defines concepts through ordering but recognizes the way in which concepts can be defined in terms of particular learners (maybe the order of the concepts changes or, more likely, their level of generality).

A related metaphor of curriculum is that of the *learning trajectory* (Breyfogle, McDuffie & Wohlhuter, 2010), which focuses on the critical moments along the path that mark progress towards a future goal. A related, student-oriented construct of *hypothetical learning trajectory* (Clements & Sarama, 2004; Confrey, Maloney, Nguyen, Mojica & Myers, 2009) has also emerged, which describes increasing levels of sophistication demonstrated by students as they work with a concept during clinical interviews and teaching experiments. While Confrey et al. (2009) acknowledge that there is not just one trajectory through the 'conceptual corridor', the metaphor of trajectory itself evokes a single, vectored direction towards a fixed node, which is presumably the attainment of a full-blown concept. The work of these scholars attends to not only a concept's sensitivity to learners, but also its sensibility to the tasks and tools involved in the learning situation. Thus, the concept becomes at least temporarily materially intertwined. However, there remains the sense that the concept is something that is acquired at the end of the trajectory, whatever that trajectory may be.

A powerful alternative approach to the role of concepts in mathematical activity can be found in the fallibilism articulated by Imre Lakatos in the early 1960s. According to Lakatos, mathematical activity is primarily about concept formation and deformation. Mathematicians are always working with ill-defined or ambiguous concepts, and as they try to prove things about those concepts, they stretch them in new directions. In his famous example, the concept of polyhedron is stretched and moulded. However, this perturbation does not occur in a vacuum; it emerges in relation to a particular problem – which, in Lakatos's case, is to determine the relation between the vertices, edges and faces of polyhedra. With Lakatos then, as with Vygotsky, concepts emerge out of particular problems and proofs, so that one can hardly talk about concepts without also talking about problems. This way of thinking about concept is much closer to that of Châtelet, not only in its connection to problems, but also in its intrinsic focus on what the concept is *doing*, that is, on the story in which it is involved.

The mathematician Paul Lockhart (2009) uses the metaphor of story to help explain the importance of context and motivation in relation to any given 'content':

> Mathematical structures, useful or not, are invented and developed within a problem context, and derive their meaning from that context. Sometimes we want one plus one to equal zero (as in so-called 'mod 2' arithmetic) and on the surface of a sphere the angles of a triangle add up to more than 180 degrees. There are no 'facts' per se; everything is relative and relational. *It is the story that matters, not just the ending.* (p. 17, *emphasis in original*)

As Dietiker (2012) argues, the metaphor of story attends much more to the ongoing, temporal unfolding of mathematical development, rather than to the culminating mathematical goals, facts or content. Furthermore, the metaphor of story suggests that there would be a different story if the sequence were changed, and that changing one part of the story would have an effect on the rest. Not only are there different paths from A to B (as with hypothetical learning trajectories), but there are also ways of getting to B that do not begin at A and, as such, it becomes even more important to focus on the problems that gave rise to A in the first place and the reciprocal relations between A and B. In line with this approach, the mathematical problem or question focusses less on which path to take to move through a given sequence of predetermined concepts and more on which stories are worth telling about mathematical events and characters.[6]

If we compare U.S. and Canadian curricula to the way that the Russian 'mathematical circles' are designed, we can see radically different conceptions of what constitutes a concept (Fomin, Genkin & Itenberg, 1996). Rather than treating concepts such as multiplication and triangles as topics, and then essentialising them in terms of some finite set of characteristics, Russian mathematical circles are organised around multi-purpose devices that help teachers and learners map the structure of mathematical relationships and invent new mathematical objects. The idea of a sequence of concepts is thus downplayed in favour of a collection of actions, which comes closer to curriculum as a story, in the sense that the focus is on *what can be done* rather than *what is*. In part, this is due to the fact that the curriculum of these mathematical circles is based on rich problem solving, but we focus here on how this approach seems to use a different notion of concept *as device*, a notion that leverages both logical and ontological aspects. As in the rest of the book, we use the term *ontological* here to refer to the ways in which mathematics is a creative and material activity that literally invents new kinds of entities. For both novices and experts, mathematics involves both realizing the possible (logical) and actualizing the virtual (ontological). If a concept is to function in both of these ways, it must sustain a certain vibrancy and vitality. In other words, a concept of this kind must be a *multi-purpose device* that resists reification while carving

[6] Ingold's (2007) distinction between transporting and wayfaring might provide a useful metaphor for thinking about the limitations of current uses of the notion of a learning trajectory. These tend to conceive of learning as the sequence of journeys one might make from one place marker to the next, a connect-the-dots approach that is quite different from the kind of trajectory that might be traced from a given initial impulse. The former involves reaching successive destinations, while the latter involves creating paths.

out new mathematical entities. Châtelet refers to these as 'allusive devices' that give rise to thought experiments 'that penetrate closer to the heart of relation and operativity' by being deliberately productive of ambiguity (p. 12). The concept is worked as though it were a kind of material medium (clay, oil, paint, ink) that lends itself to particular kinds of activity and rendering.

Consider, for instance, tapping into the potential of a concept like even/odd parity. This is precisely the kind of concept that functions simultaneously on both the logical and ontological levels.[7] Parity makes evident the structural relationship between numbers and helps the student map the terrain in which numbers live; but parity also functions importantly as a concept that engenders new kinds of mathematical objects. It marshals both logical constraints and mathematical entities, performing acts of embodiment by which new assemblages emerge. For instance, the ancient Greek proof of the irrationality of $\sqrt{2}$ that we discussed in the previous chapter relies extensively on even/odd parity. Parity is leveraged in this famous proof to unleash and make actual the virtual (the irrational number) while operating according to the laws of legitimate inference. Parity is a creative force, that is, a concept that quite literally performs an act of embodiment by which the new number is assembled; the proof brings forth the irrational number. And yet, even/odd parity simultaneously operates as a logical instrument, determining the boundary between sense and non-sense, by which the possible is realized, as we discussed in the previous chapter. Thus, the proof both brings something new into existence and attests to the truth of this affirmation. In other words, a concept of this kind is a multi-purpose device that carves out new mathematical entities. One can see how our reading of this famous proof pertains to long-standing issues in the philosophy of mathematics regarding the status of non-constructive proofs, that is, proofs that assert the existence of something without furnishing that thing. Brouwer and the intuitionists were concerned that such proofs were dangerous and were likely to lead to paradoxes. Our concern, however, is not with paradoxes, and our motive is not to banish non-constructivist proofs, but instead to analyse common practices in school mathematics and show how the ontological is always already at work.

[7] The simultaneous functioning we are describing here might evoke for readers a comparison with the process/content distinction made in the NCTM Standards. An important difference, however, is that the same concept (parity, in this case) partakes of both the content and the process. Moreover, when functioning on the ontological level, parity has much greater precision and power than process strands, such as representation, communication and visualization.

Other concepts that are used in Russian mathematical circles are divisibility, invariance, the pigeon-hole principle, the triangle inequality and graphs (networks). Readers might wonder how a principle and an inequality might be considered *concepts* on par with divisibility. Indeed, they would not appear on any scope and sequence list of concepts. But what is more important than the given list is the way the items on the list structure the mathematical experience. We want to treat these as concepts, insofar as they are put to work to actualize the virtual. Our concern will thus be to resist the downgrading of any list of concepts to a set of topics once they are taken up in practice. This will require *preserving the concept as a creative force* by ensuring that the concept is put to work in engendering new mathematical entities. Perhaps the way to ensure that the concept remains operative, mobile and creative is to focus on whether the work itself is creative. Such work would continue to be committed to bringing the new into being (a process of becoming); in other words, such work unleashes the virtual. When the concept is used only as a logical tool or as an object or relation with a finite set of properties, the ontological aspect (actualizing the virtual) is abandoned, and activity is reduced to either naming exercises (identifying instances of the concept) or applications of the concept as a rule or logical constraint. In such cases, the ossified concept fails to sustain the mobility and potentiality from which it came.

One can easily see concepts in current curricula that have been emptied of their mobility and power but might be resuscitated. One method of attack might be to look for how the curriculum is centred on theorems, given that these are considered the ultimate expressions of mathematical truth. The first step is then to excavate the concepts that are generative within these theorems. For instance, the Pythagorean theorem is often treated as no more than a tool for calculating lengths in triangles, rather than an assemblage of inventive devices that carve up matter. If a theorem is a statement that has been proven to be true based on other accepted statements and axioms within a given deductive system, then one might argue that mathematical statements become theorems if a sanctioned proof exists and if its significance warrants such stature within the system. Of course, there are well over 300 known proofs of the Pythagorean theorem, and thus it seems that the truth of the theorem is independent of the specific characteristics of its justification. Indeed, the theorem has been shown to be logically equivalent to Euclid's fifth postulate within the given Euclidean deductive system, and thus it could, in theory, function as an accepted source statement rather than a consequence of deduction. But rather than read this fact as evidence that the theorem is universally true and timeless, we follow instead the inclination to unpack the theorem for its contingency and vibrancy, and to identify the fluid concepts that act like sticky glue in keeping it together. In reconceptualizing the Pythagorean relation, we can recast its derivation as an activity that literally brings forth the squares on the sides of the triangle. This is not a matter of discovery learning or constructivism as it is conventionally conceived by reform-oriented educators. These squares are actualized, not only out of logical necessity or inductive reasoning, but also out of some material *potential* linked to the triangle. The concept of triangle is generative in this instance. And the concept of triangle is sustained through the quivering concepts of angle and line. The virtuality of each of these concepts comes to be actualized in the assemblage that is the Pythagorean theorem. In recognizing the ontological aspect of the concept, we begin to think about the squares as invented objects. The aspect of necessity attached to the logical is then opened up, with the aspect of contingency (and thus ambiguity) attached to the act of creation. We might then see how squares could have been something else, like semicircles or trapezoids, materialized on the sides of the triangle. According to this approach, the Pythagorean theorem assembles concepts that actualize various kinds of virtual shapes in the plane. And if we apply the device to three-dimensional space, it becomes a means for sculpting multi-dimensional objects. In this way, we can think of it as a concept that does more than measure and more than adhere to the rules of logic – it actualizes virtual entities.

Mathematicians have stretched the Pythagorean theorem beyond Euclidean geometry – exploring how it might operate on spherical, elliptical and hyperbolic surfaces – by redefining concepts like straight line, angle and (right) triangle to meet the new material conditions. Such generalisations are highly inventive and often demand the introduction of new concepts, as well. But we ask, for the sake of attending to the *specificity* of the Pythagorean theorem, a series of questions that pursue this process of stretching and inventing: What is left of the theorem after such stretching? How does that which remains of the theorem clearly link materially to the new geometric spaces in which it lives? How might we unpack that which persists in all these contexts? What are the ontological implications of this specific theorem? Perhaps what persists is the relationship between extension and intension, that is, the relation between the actual and the virtual as found on the edges and the squares of edges. In other words, the theorem reflects the way a continuous curve can suddenly shift direction and trap space within it, and how it does so in such a way that it simultaneously actualizes another shadow space on its other 'side'. The Pythagorean theorem shows us how the intensity that is bundled up within an extended line is set free in the second degree and how this intensity collapses problematically (and productively) at the singular point (vertex) where a radical change in direction pushes back down on the line of flight to form a closed region. Of course, this analysis seems to rob the theorem of its metric qualities, but one has to bracket them slightly even as one considers how the theorem might work in alternative geometries. This analysis also directs our attention to the particular kind of movement that is captured in the theorem and begins to show how such movement is pivotal in the shaping of the concepts at work. In attending to the flow and the trap of movement within an assemblage as it becomes a theorem, one loosens the reigns of deduction and begins to tap into the virtuality of matter. This is how we begin to honour the ontological in the logical. The Pythagorean theorem – as an ever-emergent assemblage of relations between the spaces built on various 'sides' – reveals how such acts of entrapment and their material conditions are mutually entailed.

Can we rethink other theorems within the curriculum and reinvest our concepts with ontological power? Can we recast our concepts in terms of the way they both realize the possible and actualize the virtual? And to what extent will doing so involve a shift from an emphasis on logical necessity towards an opening for contingency? In reflecting on these questions, we come to appreciate the way in which attention to ontological power can help loosen the grip of logical necessity that strongly structures our current curricula. We are not denying the role of logical necessity in mathematics, but we believe that the relationship between the logical and the ontological needs to be re-examined. Exciting new directions in mathematics curriculum might ensue if we were to question the ways (and the whys) in which the ontological is banished.

Moreover, the emergence of fuzzy and other deviant logics in the twentieth century and the fact that logicians now use strategies from algebra point to how the relationship between logic and mathematics needs to be newly studied in the context of education. Since Piaget, the mathematics curriculum has been heavily influenced by what Lundin (2011) calls a 'developmental ideology', namely, a rationale based on theories about the order and age at which children (can/should) develop knowledge of particular mathematical concepts. This ideology assumes that learning is a problem of construction and centrally involves the acquisition of logical reasoning skills; the ensuing metaphor suggests starting from the ground up and making sure the foundations are solid. Within the ideology of development, the discipline stays fixed and immobile, while the child moves slowly up, forming 'concepts' along the way.

Current theories of learning, inspired by a Vygotskian attention to sociocultural factors and, in particular, to the mediating effects of language, still often subscribe to the developmental ideology. For example, the learning trajectory approach acknowledges the role played by tasks, tools and the teacher in a child's mathematical learning, but there is still a sense in which the child is seen as scaling an edifice, acquiring knowledge of concepts along the way. Learning trajectory proponents allow for the possibility that there are multiple ways in which a learner might develop these prescribed concepts and that any description of the learning process must account for the broad environment in which the learning takes place, but as we have said before, the concepts themselves remain fixed.

We are interested in questioning the ideology of 'development'. Châtelet's insistence on the fraudulence of the concrete/abstract dichotomy invites the possibility of moving very differently in and around the 'edifice' of mathematics. In fact, the edifice image reveals itself to be a poor metaphor for a mathematics that is about virtuality and mobility, unless we are prepared to adorn mathematics with suitable devices that permit alternative modes of access and movement on and through such an edifice.

We contend that the developmental ideology draws on a particular philosophy of mathematics, one that acknowledges to some degree the contingent and sociocultural nature of knowledge, as well as the embodied nature of mathematical understanding. Nonetheless, it continues to invest in an image of mathematics that denies the materiality of the concept. This image of mathematics remains wedded to the logical – albeit broadened to embrace both the deductive and the inductive – but it refuses the ontological. Concepts are abstracted away from the physical situation, and the learner is granted only enough embodiment to perform that abstraction. Thus, the ideology of development simply sustains Aristotle's philosophical paradox, because it remains committed to an abstract/concrete binary that ultimately abandons mathematics to a realm of the inert and disembodied.

Châtelet exemplifies a materialist alternative in his examination of Grassmann's development of the concept of multiplication in the context of linear algebra (where one considers the product of vectors, rather than the product of numbers). In this context, instead of treating multiplication as repeated addition – a common trope for mathematicians and pedagogues alike – which inherits the docility of mechanistic adding-on and robs the operation of mobility and materiality, Grassmann revitalised the concept as a kind of movement. Châtelet describes how multiplication becomes 'productive of plurality' when one thinks of it as a rectangular space being carved out by the sweeping out of a line perpendicular to it, a line that has

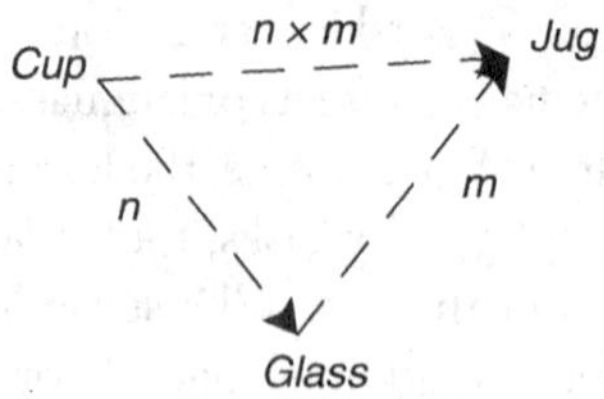

Figure 8.5. Multiplication as a change in units.

itself been produced by a moving point. The rectangle, in this case, is not simply the thing that supports intuition, nor simply that which has realized the possible, nor the mere representation of a geometric fact. Rather, it is a *mobile* unit created out of two (or more!) single extensions. In this sense, mathematics comes to matter in and of matter.

One might be reminded here of the so-called area model of multiplication which is often used in school mathematics *after* students have already learned what multiplication is (repeated addition) or, more precisely, what the solutions to a certain number of multiplication 'facts' involving natural numbers are. We see in this pedagogical ordering of concepts how the mobility of multiplication is treated as a mere illustration or representation of the concept, rather than as a force that animates the concept. One exception to this ordering of the concept can be found in the Vygotskian-inspired, measurement-based curriculum in which numbers are introduced in terms of units of measure (so that the number '3' is always actually three units of length), and multiplication is introduced in terms of a change in the system of units. For example, if we want to know how many small paper cups are needed to fill a large jug, we can determine how many small paper cups are needed to fill a larger glass (say, n), and then how many larger glasses can fill the jug (say, m). Children are taught to indicate this change of units action in the manner shown in Figure 8.5.

This multiplication-as-dilation approach, in which n becomes $n \times m$, can be shown diagrammatically to contrast with the multiplication-as-repeated-addition approach, as shown in Figure 8.6. One unit is becoming four units through a kind of cell division into equal parts. But one unit could just as easily become a half unit, and the unit itself would retain an indeterminacy. Aside from its success in developing stronger, more flexible conceptions of multiplication (see Schmittau, 2003), in comparison to the prevalent repeated addition approach, the measurement-based curriculum cultivates a pedagogy of the concept – not so much because it involves material actions (comparing and measuring objects) in this pedagogical

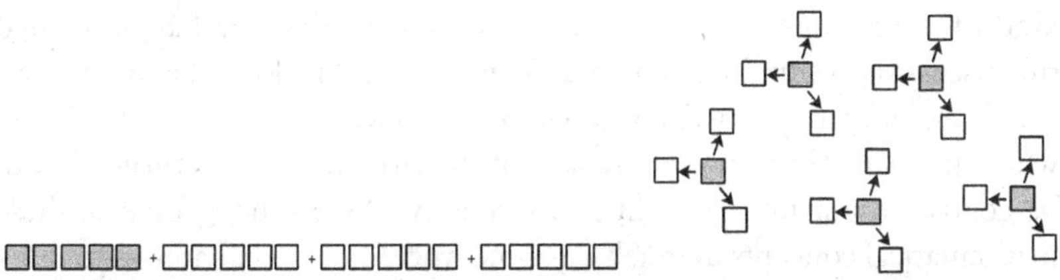

Figure 8.6. Multiplication as sequential addition or as dilation.

setting, but more because it frames multiplication as changing the nature of the objects involved (be they numbers or other entities), focussing instead on actions like dilating and equi-partitioning.

Rotman (2008) proposes an approach to notation that resonates strongly with Châtelet's technologies-driven historical studies (technologies of gestures and diagrams; perhaps the term technics is more appropriate). Rather than focussing on the typical issue of how technology will change the role of logical necessity in mathematics, Rotman provocatively claims that the advent of digital technologies will lead to new kinds of gestural and diagrammatic inventions. He anticipates the slow replacement of the alphabet – that sequential, disembodied and static mode of inscription that has dominated the Western intellectual landscape – with other kinds of sensory modalities enabled by digital interfaces. For the mathematician, this would entail moving away from a world dominated by symbolic formal language (which is also sequential, disembodied and static) and towards one experienced 'as much through touch as vision, through tactile, gestural, and haptic means' (de Kerckhove, 2006, p. 8, cited in Rotman, 2008, p. 8). Central to Rotman's argument is his assertion that mathematics has been, and will continue to be, involved in a two-way, co-evolutionary relationship with machines. If in the past, mathematicians were able to deny any machine-based agency through their discourse of detemporalization, decontextualization and depersonalization (Balacheff, 1988), the increasing presence and power of current digital technologies will assert itself. In this sense, mathematical activity will be seen more clearly as something that co-evolves with the discipline, the seeing/touching/hearing/smelling mathematician and the material world. In Rotman's work, we find a material posthumanist philosophy of mathematics that invites us to look for curricular possibilities that are less driven by developmental assumptions. This frees us to pursue a curriculum of encounters with the virtual.

In this chapter, we have drawn on Châtelet's notion of the virtual in order to develop a more robust theory of materialism that seems to suit the kind of astonishment we associate with mathematics. We have extended the discussion begun in Chapters 2, 3 and 4 about the logical and ontological facets of mathematical concepts, and we have inquired into the way in which mathematics concepts partake of the virtual. Mobility turned out to be central to our understanding of virtuality. We can now think of these mathematical concepts in much the same way that Barad invites us to think of atoms, photons, stingrays and lightning – as unfinished relational assemblages – or, as Châtelet names them, 'allusive devices'. It is important to keep in mind that the actual (the diagram, the point, the number) somehow retains the trace of this movement. In retaining this mobility, Châtelet differs significantly from other explanations of mathematical concepts that appeal to 'abstraction', because abstraction is ultimately contemptuous of the hand and the movement by which the possible was determined. Abstraction strips mathematical entities of their physical emergence and pretends that they have no experiential history. Although many scholars assume that it is this act of stripping away the contingencies of emergence that makes mathematics so effective in science, Châtelet's entire historical project is to contest this assumption and show how inventions in mathematics prove effective *precisely* because they retain a trace of the gestures (and the mobility of the virtual) that brought them forth. Thus, the virtual in sensible matter becomes intelligible, not by a reductionist abstraction or by a 'subtraction of determinations' (Aristotle's approach to abstraction), but by the actions (moving, excising, cutting through diagrams and gestures) that awaken the virtual or potential multiplicities that are always implicit in any material. These insights pertain to learning mathematics, as well, because school mathematics might be made to quiver 'with the virtualities that inhabit it', and learning assemblages might be spurred on to 'reactivate a productivity' that congeals and stiffens too quickly after its articulation (Châtelet, 1993/2000, p. 103).

Conclusion

This book takes up and transforms the human body in ways that might make it unrecognizable to many readers. Can we really suggest that the body be reconceived as a heterogeneous assemblage of organic matter, concepts, sound and software? Can it be considered an assemblage always in the process of becoming, always growing, contracting and mutating? If at first there seems to be little gained from such radical reconfiguring, we hope that our attempts to delve into the implications of this approach have proven thought-provoking and insightful. By looking for when and how a body comes to be recognized as a body, we have tried to study the intricate unfolding of mathematical activity, to track the processes of individuation by which learners and concepts become enjoined. We have argued that the mathematical body comes into being (is always *becoming*) as an assemblage of trembling relations. Assemblages can be thought of in terms of knots and meshworks, which evoke less an image of different bits and pieces glued to each other than an image of a tangled set of paths, each with its own mobility and degree of freedom. The notion of entanglement refers both to the impossibility – despite common parlance – of isolating any one part of an assemblage as self-sustaining or ontologically distinct and to the relational forces of the paths as they vibrate with intensity, quaking and quivering with potentiality. This potentiality of the body is linked to its indeterminate form, or as Michel Serres suggests, its 'multiple postures':

> How do we define a body given over to so many poses and signs: when and under which form is it itself? How do we get beyond so many differences according to the person: when and under which form is it us? These multiple postures prevent us from saying. My body and our species don't exist so much in concrete reality as 'in potency' or virtuality. (Serres, 2011, p. 52)

One of the challenges of this book has been to show how the virtuality to which Serres alludes is also embodied in mathematical concepts, insofar as they are forceful, animate, mobile, alive and material. We used the term *becoming mathematics* to explore the ways that both mathematics and the doer of mathematics are in motion. That is to say, both the subject and object of learning are in the midst of an ontological process of change or alteration, each moving away from that which they were and towards something entirely new. Becoming mathematics in the context of the classroom, then, pertains to the development of a *minor* mathematics, a mathematics that is not the state-sanctioned discourse of school mathematics but that might be full of surprises, non-sense and paradox.[1] As a kind of school mathematics, it might lack scalability, cross-country comparability and within-country standardization, and thus be at odds with current institutional demands. However, a minor mathematics is likely to engage students and teachers in more expansive ways, and our hope is that it would engage *more* students in mathematics.

With the help of Châtelet and his concept of the virtual, we were able to put forward the notion of the mathematical body – thus stretching the notion of the body to include not only the human one or the physical one (of tools and objects), but also the body of the concept. While many readers can accept the proposal that physical objects are material, such a proposal in the case of the body of mathematics challenges well-entrenched assumptions about the abstract, immaterial nature of mathematical concepts. Our goal was to show how Châtelet's sense of the virtual could rematerialize mathematical concepts without robbing them of their power and potential for fantasy. The virtual is that which partakes of the real, thus enabling the bridge between mathematical and physical worlds. But this mathematical world is not an abstract or detached world ground only in logical deduction and the derivation of what is possible. The bridge is not a correlation between ontologically distinct worlds, nor is it only an epistemic concern. In that sense, the virtual is not just about logical possibility, which might be completely divorced from the real. The virtual in sensible matter becomes intelligible not by a reductionist abstraction, but by the actions (diagrams/gestures/spoken words/caresses) that awaken the virtual or potential multiplicities that are entailed in any surface or material site. The virtual also pertains to the indeterminacy at the source of all actions and all matter,

[1] See Deleuze and Guattari (1987) for a discussion of how minor sciences and literatures are powerful forces that operate alongside state-sanctioned policy and practice. Also see de Freitas 2012c for how this relates to mathematics classroom discourse.

whereas the possible pertains to the compliance of our actions with logical and perceptual constraints. Given the centrality of the virtual in Châtelet's work, we wanted to see how it might apply to school mathematics; we wanted to see how the concepts that drive curricula might be made to matter. We explored the virtuality of a range of concepts in an attempt to span the school curriculum, seeking ways to mobilize familiar, and often inert, concepts. This led us to Cutler and MacKenzie's notion of the 'pedagogy of the concept', in which concepts are revised to become material and ontogenetic devices. While our virtualizing of concepts might be seen as an epistemological project, we argued that it is also an ontological project and a didactical one, in that we can think of learning-teaching as that which sustains the mobility of concepts. We argued that such a pedagogy might function to leverage both the logical and ontological aspects of concepts, sustaining student awareness of the virtuality and indeterminism at the source of curricular concepts. And yet, we have not offered any programs for instruction or any lesson plans. We hope, however, to have outlined how a very different philosophical approach to concepts would radically alter our understanding of curriculum and our teaching practices.

Châtelet's historical, material analyses inspired us to try similarly radical readings of classroom mathematical activity. In Chapter 3, we unpacked the gestural nature of student diagrams, as well as the particular assemblages of student-circle-diagram that emerged. We identified places of invention where, for example, students added a dotted line to the paper, thereby bringing into being a new dimension, or where they marked up the page with a series of arrows, thereby forging new temporal relationships between such entities. In Chapter 4, we were able to study the inventiveness of primary school mathematics activity in terms of the new boundaries enabled by various digital devices. Châtelet's material readings informed each of our video analyses, including alternate interpretations of the classroom videos in Chapters 5 and 7 that had been analysed previously by other researchers. Each of these analyses also aimed to highlight new insights from material understandings of language and (dis)ability, respectively. In all of these, we focussed on the way classroom mathematical activity might constitute a place of invention and discovery, where learning mathematics is conceived in terms of generative and creative acts that carve up space and evoke new dimensions and new structures.

We characterized these acts as actualizations of the virtual dimensions of matter. The examples in Chapter 3 focussed specifically on the diagrams produced by learners, working as we were in the spirit of Châtelet. However, in Chapter 4, we took a broader unit of analysis that involved much more than diagrams. Here we proposed a definition of inventive acts that borrowed from Châtelet but sought to refine an appropriate inventiveness in the classroom context. We extended Elizabeth Grosz's description of freedom to our own goals, developing a four-fold characterization of inventive acts that aimed to capture both the ontological and social semiotic facets of creativity. We argued that creative acts should introduce or catalyse the new – quite literally bringing forth or making visible that which was not present earlier; they should not align with current habits and norms of behaviour; be unexpected or unscripted, in other words, without prior determination or direct cause; or be without *given* content, in that the meaning should not be exhausted by existent meanings. The examples we analysed throughout the book, but especially those in Chapters 2, 4 and 8, were meant to show that the process of actualization – the act of inventive mathematical activity – is a material act, whereby the new is generated and the ontological bond between the actual/virtual is remade.

Our analyses of these classroom videos addressed the classroom environment as a whole, rather than the capacities or abilities of individual students. In their classroom-level focus, these analyses echoed the flavour of Yackel and Cobb's (1996) study of socio-mathematical norms, which attends to the development and deployment of norms about what counts as being mathematically different, sophisticated, efficient and elegant. Our analyses differ in part from those of our colleagues, because we insist on the way meanings are assemblages involving gestures, drawings, movements and technologies, as well as ways of talking. We focus less on how individuals develop mathematical ways of talking than on how inventiveness emerges out of particular assemblages involving various kinds of gestural relationships. That said, in Chapter 5, we turned our attention very specifically to 'ways of talking' in an attempt to pursue the question of how a material understanding of language might help advance the already insightful discursive approaches to research in mathematics education. We focussed specifically on the materiality of language, which we located not only in terms of the physicality of speech (the movement of the throat, mouth and lungs, the sound of the voice), but also in the phenomenon of sound in the classroom (how it moves, pushes, embraces and silences). We showed how our material approach changes the way we might hear a classroom video without simply listening for the meaning of words. Such a way of hearing helped us decentre individual speakers (without, of course, ignoring them), in order to listen for the way sound partakes in the forming of learning-teaching assemblages. We described this kind of hearing in terms of modulation, which captures the idea that classroom discourse involves a block of sensation collectively intensifying. We defined language use in terms of in(ter)vention, rather than communication, in an attempt to do justice to the very materiality of language, particularly in terms of the prosody of speech, but also as a means to study the way meaning is bound up with the perceptual contingencies of an assemblage.

We borrowed Manning's use of the term 'affective tonality' to describe the way intensity flows across spoken words, at times independent of, and possibly at odds with, the intent of communication. Thus, we used the term 'affect' to designate the materiality of speech and the mobility and flexibility of the human body. Here we followed Sheets-Johnstone's (2009) emphasis on the affective/kinetic dynamic, in which affectivity involves the moving towards or away from other animate and inanimate bodies. Her non-dualistic approach echoes Spinoza's conception of affect, which challenges the idea that affective states are an interior experience, reflected in externally visible behaviour. This is how Spinoza can assert that '[t]here is no longer a subject' but instead 'only individuating affective states of an anonymous force' (Spinoza, cited in Alliez, 2004, p. 27).[2] Deleuze and Guattari (1987) refine Spinoza's work by defining affect as 'a prepersonal intensity corresponding to the passage from one experiential state of the body to another and implying an augmentation or diminution in that body's capacity to act' (p. xvi). Affect is thus not synonymous with feeling or emotion as a state of being, but is strongly tied to the body's capacity to affect and be affected. Affect is relational and contingent, as well as dispersed. This take on affect is dissociated from studies that consider affect as part of a representational domain or that seek to determine how affect is distinct from cognition. Our approach treats affect as a publically circulating energy, a modulating intensity saturating the assemblage – the way enthusiasm or boredom moves across a classroom or the way waves of anger or joy reverberate in a conversation. Thus, our aim is to study affect as part of the intra-action of mathematical activity. By attending to the modulation of intensity, we move away from typologies of affect and emotion, which are all too often assumed to be possessed by the individual, and towards the study of affect as a collectively enacted form of engagement. We believe that our approach to studying affect opens up new ways of researching classrooms and new ways of exploring the affective tonality of learning-teaching more generally.

[2] This contrasts with Vygotskian-inspired work that emphasizes the relation between affect and cognition in terms of social and cultural practices, which always has to be mediated (see Walshaw & Brown, 2012).

By attending carefully to the details of sensory activity in mathematics, such as drawing a dotted line, repeatedly tapping a ruler on a number line or the prosodic inflection of a voice, we have showed how mathematical engagement operates on various material planes and that all of these planes are folds in the socio-political fabric of life. The convergence of sensation with *common* sense marks the entrenchment of a form of life – perhaps also a regime of truth – in which certain senses become entrained and regulated as part of curriculum and policy. Forms of common sense are validated through material classroom practices – what Barad calls boundary-making practices – that delineate an inside and an outside, access and denial, success and failure. In Chapter 7, we mapped the boundary-making practices at work in mathematics, where aesthetic/political acts are described in terms of the disrupting of given boundaries that designate sense and non-sense. In Chapter 6, we also unpacked boundary-making practices that produce (dis)ability in mathematics education, while simultaneously regulating the limits of the human body and the configuration of the sense organs.

Rancière's notion of the 'distribution of the sensible' and the paradoxical autonomy of the aesthetic object proved very useful in developing an inclusive materialism relevant to mathematics. We examined how this kind of autonomy operates within the history of mathematics, where the aesthetic – as both sensory knowing and value judgement – plays a constitutive, if often covert, role. Although our historical analysis was focussed only on some key Western developments in mathematics, we analysed various instances of consensus that have emerged within the field in an effort to explore the political dimension of the mathematical aesthetic. We then turned our attention to the policies of mathematics education to show how particular ways of doing and making that have been expounded in these policies entail certain distributions of the sensible that inevitably produce regulative – and increasingly invasive – consensus. Mobilizing Rancière's notion of dissensus, we explored the kind of political acts that might make visible and audible current assumptions about what *makes sense*, what is intelligible and, ultimately, what is mathematical. If consensus is an alignment between sense (as sensation) and sense (as meaning), then dissensus is that which breaks up this alignment. Our exploration of consensus and dissensus in mathematics (as well as school mathematics) pointed to the importance of such mathematical practices as monster-making experiences of surprise.

The philosophical framework within which we work – inclusive materialism – is part of a paradigm shift in the social sciences more generally. This shift has been slow in coming to studies of mathematics – perhaps because of the perennial challenges of thinking about the mathematical concept and matter together (Hacking, 2012) – but the shift has been steadily growing in other domains. For instance, the 'material turn' in anthropology, along with 'thing theory' in literary and cultural studies and the 'non-human turn' of the arts, humanities and social sciences, all point to a broad and resurgent interest in the murky boundaries between people and objects. We chose to trace the emergence of new materialisms primarily from within feminist philosophy, in part because the question of the body has always been central and complex in this intellectual tradition. Despite this motivating thread, throughout the book we have remained silent on the issue of how bodies are gendered. In Chapters 6 and 7, we wrote about the potentiality of the body as a political issue, pointing to how configurations of the senses were coupled to particular valorizations of a mathematical aesthetic. We showed how this kind of coupling was taken up in curriculum policy and how it operated alongside particular pedagogical practices, but we did not speak to the gendering of mathematics, nor to the gendered identities of students and teachers in mathematics classrooms. It would seem to be an oversight, given the accomplishments made by feminist science studies over the last three decades in demonstrating how knowledge paradigms and practices are performative of gender. From a post-humanist perspective, however, the question of gender has to be entirely recast so as to consider non-human relations in ways that do not simply reduce these relations to projections of human accounts of gender. And at the same time, one has to be sure not to reduce the substance of these accounts to biological notions of sex, because this is in part what many waves of feminism have taught us. In keeping this tension at the fore, a post-humanist perspective cannot deny the ubiquity of sex as a fundamental divider of biological populations, both human and non-human. The complexity of tackling this issue – how it demands careful examination of sex and acts of gendering in terms that stretch beyond the human – has only recently been *felt* in post-humanist scholarship on the anthropocene period, which is the geologic term used to designate the period 'after man' (Colebrook, 2013). Given the ambitions of our book and our focus on the materiality of mathematical concepts, we did not feel that we could possibly do justice to this topic, but we recognize the need to address it, and we encourage other researchers to join us in exploring the complex ways in which bodies are configured according to prevalent tendencies, like gender.

We hope that one of the distinctive contributions of this book has been to show educational researchers how they can leverage insights from the history and philosophy of mathematics in order to study mathematics teaching

MATERIALIST TURN IN SUBJECTS

and learning. In particular, we have drawn extensively on Châtelet's study of inventiveness in the history of mathematics (such as Hamilton's quaternions, Grassmann's theory of extension and Cauchy's residue theorem) to offer a radically new perspective on the nature of student inventiveness. We also tapped the history and philosophy of mathematics to show how different communities of mathematical sense-making emerged at different times and locations. Drawing also on the writings of mathematicians such as Hardy, Poincaré, Rota and Lockhart, we showed how the mathematic aesthetic operates by insisting on the conjunction of two oppositional concepts of sense, the first associated with the autonomy of expression and the second with the enactment of a common form of sensibility. Similarly, we were able to show how different configurations of the senses have been validated and associated with mathematical prowess in different ways over time. We offered glimpses of this throughout history, where mathematical ability has been defined and redefined in terms of sensory (dis)abilities.

Drawing on recent critical (dis)ability theory, we continued to challenge our understanding of the human body. In particular, we suggested that a post-humanist pedagogy might help us move beyond Enlightenment notions of mathematical ability by better studying the collective contingency of perception as it unfolds in classrooms. The changing historical relationship between the senses and mathematical (dis)ability is an area that needs much more research, but our aim here has been to show how inclusive materialism allows us to map these contingent and variable relationships and to raise questions that demand we examine our assumptions about mathematics achievement. Indeed, Rancière's notion of communities of sense makes all too obvious the contingency of mathematics as a changing aesthetic practice of making and doing. We are not naïve enough to think that determinations regarding ability will, or even can, go away; however, our commitment to the works of scholars like Deleuze, Châtelet and Rancière involves a certain orientation towards the inventive, the new and the future – preparing the way for a people yet to come.

Bibliography

Ahlquist, R. (2001). Critical multicultural mathematics curriculum: Multiple connections through the lenses of race, ethnicity, gender, and social class. In J. E. Jacobs, J. R. Becker, & G. F. Glimer (Eds.), *Changing the faces of mathematics: Perspectives on gender* (pp. 25–36). Reston, VA: NCTM Publishing.

Adler, J. (2001). *Teaching mathematics in multilingual classrooms*. Dordrecht, The Netherlands: Kluwer.

Alibali, M. W., & Nathan, M. J. (2012). Embodiment in mathematics teaching and learning: Evidence from students' and teachers' gestures. *Journal of the Learning Sciences, 21*(2), 247–286.

Alliez, É. (2004). *The signature of the world: What is Deleuze and Guattari's philosophy?* London: Continuum.

Apple, M. (1999). *Power, meaning, and identity: Essays in critical educational studies*. New York: Peter Lang.

Applebaum, P. (1999). Target: Number. In S. R. Steinberg, J. L. Kincheloe, & P. H. Hinchey (Eds.), *The post-formal reader: Cognition and education* (pp. 423–448). London: Falmer Press.

Aristotle. (1986). *De anima* [On the soul]. (H. Lawson-Trancred, Trans.). New York: Penguin.

Asperger, H. (1944/1991) 'Autistic psychopathy' in childhood. In U. Frith (Ed.), *Autism and Asperger Syndrome* (pp. 37–92). Cambridge: Cambridge University Press. (Original work published 1944)

(1944). Die 'autistischen Psychopathen' im Kindesalter. *Archiv für Psychiatrie und Nervenkrankheiten, 117*, 76–136.

Atweh, B., Bleicher, R. E., & Cooper, T. J. (1998). The construction of the social context of mathematics classrooms: A sociolinguistic analysis. *Journal for Research in Mathematics Education, 29*(1), 63–82.

Badiou, A. (2008). *Number and numbers*. New York: Polity Press.

Bakhtin, M. M. (1952/1986). The problem of speech genres. (V. McGee, Trans.). In C. Emerson & M. Holquist (Eds.), *Speech genres and other late essays* (pp. 60–102). Austin: The University of Texas Press. (Original work published 1952)

Bakker, A., & Hoffmann, M. H. G. (2005). Diagrammatic reasoning as the basis for developing concepts: A semiotic analysis of students' learning about statistical distribution. *Educational Studies in Mathematics, 60*, 333–358.

Balacheff, N. (1988). Aspects of proof in pupils' practice of school mathematics. In D. Pimm (Ed.), *Mathematics, teachers and children* (pp. 216–235). London: Hodder and Stoughton.

Ball, D. L. (1993). With an eye on the mathematical horizon: Dilemmas of teaching elementary school mathematics. *Elementary School Journal*, *93*(4), 373–397.

Ball, D. L., & Cohen, D. K. (1996). Reform by the book: What is – or might be – the role of curriculum materials in teacher learning and instructional reform? *Educational Researcher*, *25*, 6–14.

Barad, K. (2003). Posthumanist performativity: How matter comes to matter. *Signs*, *28*(3), 801–831.

(2007). *Meeting the universe halfway: Quantum physics and the entanglement of matter and meaning*. Durham, NC: Duke University Press.

(2008). Living in a posthumanist material world: Lessons from Schrodinger's cat. In A. Smelik & N. Lykke (Eds.), *Bits of life: Feminism at the intersections of media, bioscience, and technology* (pp. 165–176). Seattle: University of Washington Press.

(2010). Quantum entanglements and hauntological relations of inheritance: dis/continuities, spacetime enfoldings, and justice-to-come. *Derrida Today*, *3*(2), 240–268.

(2011). Nature's queer performativity. *Critical Humanities and Social Sciences*, *19*(2), 121–158.

Baron-Cohen, S. (2001). Theory of mind in normal development and autism. *Prisme*, *34*, 174–183.

Baron-Cohen, S., Wheelwright, S., Skinner, R., Martin, J., & Clubley, L. (2001). The autism-spectrum quotient (AQ): Evidence from Asperger Syndrome/high-functioning autism, males and females, scientists and mathematicians. *Journal of Autism and Developmental Disorders*, *31*, 5–17.

Baron-Cohen, S., Wheelwright, S., Stone, V., & Rutherford, M. (1999). A mathematician, a physicist and a computer scientist with Asperger Syndrome. *Neurocase*, *5*, 475–483.

Bartolini-Bussi, M., & Boni, M. (2003). Instruments for semiotic mediation in primary school classrooms. *For the Learning of Mathematics*, *23*(2), 12–19.

Bateson, G. (1972). *Steps to an ecology of mind: Collected essays in anthropology, psychiatry, evolution, and epistemology*. Chicago: University of Chicago Press.

Baxter, J. A., Woodward, J., Voorhies, J., & Wong, J. (2002). We talk about it, but do they get it? *Learning Disabilities Research & Practice*, *17*, 173–185.

Becker, A. L. (1995). *Beyond translation: Essays toward a modern philology*. Ann Arbor: University of Michigan Press.

Benbow, C. B. (1987). Possible biological correlates of precocious mathematical reasoning ability. *Trends in the Neurosciences*, *10*, 17–20.

Bennett, J. (2010). *Vibrant matter: A political ecology of things*. Durham, NC: Duke University Press.

Benwell, B., & Stokoe, E. H. (2006). *Discourse and identity*. Edinburgh: Edinburgh University Press.

Bernstein, B. (1996). *Pedagogy, symbolic control and identity: Theory, research, critique*. London: Taylor & Francis.

Bhabha, H. (1994). *The location of culture*. London: Routledge.

(2003). A statement for the Critical Inquiry board. *Critical Inquiry*, *30*(2).

Bishop, A. J. (1988). *Mathematical enculturation: A cultural perspective on mathematics education*. Dordrecht, The Netherlands: Kluwer.

Boaler, J., & Humphreys, C. (2005). *Connecting mathematical ideas: Middle school cases of teaching and learning*. Portsmouth, NH: Heinemann.

Bochicchio, D., Cole, S., Ostein, D., Rodriguez, V., Staples, M., Susia, P., & Truxan, M. (2009). Shared language. *Mathematics Teacher*, *102*(8), 606–613.

Boi, L. (2004). Questions regarding Husserlian geometry and phenomenology. A study of the concept of manifold and spatial perception. *Husserl Studies*, *20*, 207–267.

Boler, M., & Zembylas, M. (2003). Discomforting truths: The emotional terrain of understanding difference. In P. P. Trifonas (Ed.), *Pedagogies of difference: Rethinking education for social change* (pp. 110–136). New York: Routledge-Falmer.

Borgioli, G. M. (2008). A critical examination of learning disabilities in mathematics: Applying the lens of ableism. *Journal of Thought*, 43(12), 131–147.

Bostock, D. (2009). *The philosophy of mathematics: An introduction*. Oxford: Wiley-Blackwell.

Bourdieu, P., & Passeron, J. C. (1990). *Reproduction in education, society and culture*. Thousand Oaks, CA: Sage Publications.

Brantlinger, E. (2001). Poverty, class, and disability: A historical, social, and political perspective. *Focus on Exceptional Children*, *33*(7), 1–19.

(2004). Confounding the needs and confronting the norms: An extension of Reid and Valle's essay. *Journal of Learning Disabilities*, *37*(6), 490–499.

Bremigan, E. G. (2001). Dynamic diagrams. *Mathematics Teacher*, *94*(7), 566–574.

(2005). An analysis of diagram modification and construction in students' solutions to applied calculus problems. *Journal of Research in Mathematics Education*, *36*(3), 248–277.

Breyfogle, M. L., McDuffie, A. M., & Wohlhuter, K. A. (2010). Developing curricular reasoning for grades preK–12 mathematics instruction. In B. Reys, R. E. Reys, & R. Rubenstein (Eds.), *Mathematics curriculum: Issues, trends, and future directions. NCTM 2010 Yearbook* (pp. 307–320). Reston, VA: NCTM Publishing.

Briggs, C. L. (1993). Metadiscursive practices and scholarly authority in folkloristics. *Journal of American Folklore*, *106*(422), 387–434.

Brouwer, L. E. J. (1952). Historical background, principles and methods of intuitionism. *South African Journal of Science*, *49*, 139–146.

Brown, J. R. (2008). *Philosophy of mathematics: A contemporary introduction to the world of proofs and pictures* (2nd ed.). New York: Routledge.

Brown, T., & McNamara, O. (2011). *Becoming a mathematics teacher: Identity and identifications*. New York: Springer.

Brown, S. I., & Walter, M. I. (1983). *The art of problem posing*. Philadelphia: Franklin Institute Press.

Burbules, N. C. (2006). Rethinking the virtual. In J. Weiss, J. Nolan & P. Trifonas (Eds.), *The international handbook of virtual learning environments* (pp. 37–58). Dordrecht: The Netherlands: Springer.

Burton, L. (1999). Why is intuition so important to mathematicians but missing from mathematics education? *For the Learning of Mathematics*, 19(3), 27–32.

Butler, J. (1993). *Bodies that matter*. New York: Routledge.

(1997). *The psychic life of power: Theories in subjection*. Stanford, CA: Stanford University Press.

Cavaillès, J. (1970). Logic and the theory of science. In J. J. Kockelmans & T. J. Kisiel (Eds.), *Phenomenology and the natural sciences* (pp. 353–409). Evanston, IL: Northwestern University Press.

Ceglowski, D. (2002). Research as relationship. In D. K. Denzin & Y. S. Lincoln (Eds.), *The qualitative inquiry reader* (pp. 5–24). Thousand Oaks, CA: Jossey-Bass.

Châtelet, G. (1987). L'enchantement du virtuel. *Chimères*, *2*. Retrieved from http://www.revue-chimeres.fr/drupal_chimeres/?q=node/16

(1993/2000). *Figuring space: Philosophy, mathematics, and physics*. (R. Shore & M. Zagha, Trans.). Dordrecht, The Netherlands: Kluwer. (Original work published as *Les enjeux du mobile*, 1993)

(2006). Interlacing the singularity, the diagram and the metaphor. In S. B. Duffy (Ed. and Trans.), *Virtual mathematics: The logic of difference* (pp. 31–45). Manchester: Clinamen.

Cheah, P. (2010). Non-dialectical materialism. In D. Coole & S. Frost (Eds.), *New materialisms: Ontology, agency, and politics* (pp. 70–91). London: Duke University Press.

Christensen, O. R., Stentoft, D., & Valero, P. (2008). Power distributions in the network of mathematics education practices. In E. de Freitas & K. Nolan (Eds.), *Opening the research text: Critical insights and in(ter)ventions into mathematics* (pp. 147–154). New York: Springer.

Christie, F., & Martin, J. R. (Eds.). (2007). *Language, knowledge and pedagogy: Linguistic and sociological perspectives*. London: Continuum International Publishing Group.

Clagett, M. (1968). *Nicole Oresme and the medieval geometry of qualities and motions*. Madison: The University of Wisconsin Press.

Clandinin, D. J., & Connelly, F. M. (1995). *Teacher's professional knowledge landscapes*. New York: Teachers College Press.

(2000). *Narrative inquiry: Experience and story in qualitative research*. San Francisco: Jossey-Bass.

Clements, D. H., & Sarama, J. (Eds.). (2004). Hypothetical learning trajectories [Special issue]. *Mathematical Thinking and Learning*, *6*(2), 163–184.

Clough, P. (2002). *Narratives and fictions in educational research*. Philadelphia: Open University Press.

Coates, J. (1996). *Women talk*. Oxford: Blackwell.

(2003). *Men talk*. Oxford: Blackwell.

Cole, A. L., & Knowles, J. G. (2003). *Provoked by art: Theorizing arts-informed inquiry*. Lanham, MD: Rowman & Littlefield.

Colebrook, C. (2008). Bourgeois thermodynamics. In I. Buchanan & N. Thoburn (Eds.), *Deleuze and politics* (pp.121–138). Edinburgh: Edinburgh University Press.

(2013). Modernism without women: The refusal of becoming-woman (and post-feminism). *Deleuze Studies*, 7(4), 427–455.

Combes, M. (2013). *Gilbert Simondon and the philosophy of the transindividual.* (T. Lamarre, Trans.). Cambridge, MA: MIT Press.

Confrey, J., Maloney, A., Nguyen, K., Mojica, G., & Myers, M. (2009). Equipartitioning/splitting as a foundation of rational number reasoning. In M. Tzekaki, M. Kaldrimidou, & C. Sakonidis (Eds.), *Proceedings of the 33rd Conference of the International Group for the Psychology of Mathematics Education (PME 33)* (Vol. 1, pp. 345–352). Thessaloniki, Greece: PME.

Coole, D., & Frost, S. (Eds.). (2010). *New materialisms: Ontology, agency, and politics.* London: Duke University Press.

Coxeter, H. S. M. (1974). *Projective geometry* (2nd ed.). Toronto: University of Toronto Press.

Crary, J. (1990). *Techniques of the observer: On vision and modernity in the nineteenth century*. Cambridge, MA: MIT Press.

Crowe, M. (1985). *A history of vector analysis: The evolution of the idea of a vectorial system.* New York: Dover Publications.

Cutler, A., & MacKenzie, I. (2011). Bodies of learning. In L. Guillaume & J. Hughes (Eds.), *Deleuze and the body* (pp. 53–72). Edinburgh: Edinburgh University Press.

Davis, B., & Simmt, E. (2003). Understanding learning systems: Mathematics education and complexity science. *Journal for Research in Mathematics Education, 34*(2), 137–167.

Davis, G. A., & Rimm, S. B. (2004). *Education of the gifted and talented* (5th ed.). Boston: Pearson Education.

de Freitas, E. (2004a). Pre-service teachers and the re-inscription of whiteness: Disrupting dominant cultural codes through textual analysis. *Teaching Education, 16*(2), 151–164

(2004b). Plotting intersections along the political axis: The interior voice of dissenting mathematics teachers. *Educational Studies in Mathematics, 55*, 259–274.

(2008a). Troubling teacher identity: Preparing mathematics teachers to teach for diversity. *Teaching Education, 19*(1), 43–55.

(2008b). Enacting identity through narrative: Interrupting the procedural discourse in mathematics classrooms. In T. Brown (Ed.), *The psychology of mathematics education: A psychoanalytic displacement* (pp. 139–155). Rotterdam, The Netherlands: Sense Publishers.

(2010). Making mathematics public: Aesthetics as the distribution of the sensible. *Educational Insights, 13*(1). Retrieved from http://www.ccfi.educ.ubc.ca/publication/insights/v13n01/articles/defreitas/index.html

(2012a). The classroom as rhizome: New strategies for diagramming knotted interaction. *Qualitative Inquiry, 18*(7), 557–570.

(2012b). The diagram as story: Unfolding the event-structure of the mathematical diagram. *For the Learning of Mathematics, 32*(2), 27–33

(2012c). What were you thinking? A Deleuzian/Guattarian analysis of communication in the mathematics classroom. *Educational Philosophy and Theory, 45*(3), 287–300.

(2013). The mathematical event: Mapping the axiomatic and the problematic in school mathematics. *Studies in Philosophy and Education, 32*(6), 581–599.

de Freitas, E., & McCarthy, M. (2012). (Dis)orientation and spatial sense: Topological thinking in the middle grades. Paper presented at CERME 8, Working Group 4: Geometrical Thinking, Manavgat-Side, Antalya–Turkey.

de Freitas, E., & Sinclair, N. (2012). Diagram, gesture, agency: Theorizing embodiment in the mathematics classroom. *Educational Studies in Mathematics, 80*, 133–152.

de Freitas, E. & Zolkower, B. (2009). Using social semiotics to prepare mathematics teachers to teach for social justice. *Journal of Mathematics Teacher Education, 12*(3), 187–203.

de Freitas, E., & Zolkower, B. (2011). Developing teacher capacity to explore non-routine problems through a focus on the social semiotics of mathematics classroom discourse. *Research in Mathematics Education, 13*(3), 229–247.

Delanda, M. (2006). *A new philosophy of society: Assemblage theory and social complexity*. London & New York: Continuum.

(2008). Deleuze, materialism and politics. In I. Buchanan & N. Thoburn (Eds.), *Deleuze and politics* (pp. 160–177). Edinburgh: Edinburgh University Press.

Deleuze, G. (1984). *Kant's critical philosophy: The doctrine of the faculties*. Minneapolis: University of Minnesota Press.

(1990). *The logic of sense*. (M. Lester, Trans.). New York: Columbia University Press.

(1993). *The fold: Leibniz and the Baroque*. Minneapolis: Regents of the University of Minnesota Press.

(1994). *Difference and repetition*. (P. Patton, Trans.). New York: Columbia University Press.

(2001). *Pure immanence: Essays on a life*. (J. Rajchman, Trans.). New York: Zone Books.

(2003). *Francis Bacon: The logic of sensation*. (D. W. Smith, Trans.). Minneapolis: University of Minnesota Press.

Deleuze, G., & Guattari, F. (1987). *A thousand plateaus: Capitalism and schizophrenia*. Minneapolis: University of Minnesota Press.

Derrida, J. (1974). *Of grammatology*. Baltimore: Johns Hopkins University Press.

Dewey, J. (1902). The child and the curriculum. In L. A. Hickman & T. M. Alexander (Eds.), *The essential Dewey* (Vol. 1, pp. 236–245). Chicago: University of Chicago Press.

Dietiker, L. (2012). *The mathematics textbook as story: A literary approach to interrogating mathematics curriculum*. (Unpublished PhD dissertation). Michigan State University, East Lansing, MI.

Diezmann, C. M., & English, L. D. (2001). Promoting the use of diagrams as tools for thinking. In A. A. Cuoco (Ed.), *2001 National Council of Teachers of Mathematics yearbook: The role of representation in school mathematics* (pp. 77–89). Reston, VA: NCTM Publishing.

Dowling, P. (1998). *The sociology of mathematics education: Mathematical myths/pedagogic texts*. London: The Falmer Press.

Doxiadis, A. (2003, January 3). Embedding mathematics in the soul: Narrative as a force in mathematics education. *Opening address to the Third Mediterranean Conference on Mathematics Education*, Athens. Retrieved from http://www.apostolosdoxiadis.com/files/essays/embeddingmath.pdf

(2004, June 11). *The mystery of the black knight's Noetherian ring*. Keynote address at the Fields Symposium on online mathematical investigation as a narrative experience, University of Western Ontario.

Doxiadis, A., & Mazur, B. (Eds.). (2011). *Circles disturbed: The interplay of mathematics and narrative*. Princeton: Princeton University Press.

Drake, C., Spillane, J. P., & Hufferd-Ackles, K. (2001). Storied identities: Teacher learning and subject matter context. *Journal of Curriculum Studies, 33*(1), 1–23.

Dreyfus, T., & Eisenberg, T. (1986). On the aesthetics of mathematical thought. *For the Learning of Mathematics, 6*(1), 2–10.

Driscoll, M. (2003). The sound of problem solving. In H. L. Schoen & R. I. Charles (Eds.), *Teaching mathematics through problem solving: Grades 6–12* (pp. 161–175). Reston, VA: NCTM Publishing.

Dunn, T. K. (2005). Engaging prospective teachers in critical reflection: Facilitating a disposition to teach mathematics for diversity. In A. Rodriguez & R. Kitchen (Eds.), *Preparing mathematics and science teachers for diverse classrooms: Promising strategies for transformative pedagogy* (pp. 143–158). Mahwah, NJ: Lawrence Erlbaum.

Dupuis, J. (2012). *Gilles Deleuze, Félix Guattari et Gilles Châtelet*. Paris: Harmattan.

Dürrenmatt, F. (1989). *Durcheinandertal*. Zürich: Diogenes.

Eide, B. L., & Eide, F. F. (2011). *The dyslexic advantage: Unlocking the hidden potential of the dyslexic brain*. New York: Penguin.

Ellis, C. (2004). *The ethnographic I: A methodological novel about autoethnography*. Lanham, MD: Altamira Press.

Ellis, C., & Bochner, A. P. (2000). Auto-ethnographic inquiry. In N. K. Denzin & Y. S. Lincoln (Eds.), *The handbook of qualitative research* (pp. 733–768). Thousand Oaks, CA: Sage Publications.

(2002). *Ethnographically speaking*. Lanham, MD: Altamira Press.

Ernest, P. (1991). *The philosophy of mathematics education*. London: Routledge-Falmer.

(1998). *Social constructivism as a philosophy of mathematics*. Albany, NY: State University of New York Press.

Fairclough, N. (1993). Linguistic and intertextual analysis within discourse analysis. In A. Jaworski & N. Coupland (Eds.), *The discourse reader* (pp. 183–212). New York: Routledge.

(2003). *Analyzing discourse: Textual analysis for social research*. New York: Routledge.

Felman, S. (1990). *Jacques Lacan and the adventure of insight: Psychoanalysis in contemporary culture*. Cambridge, MA: Harvard University Press.

Ferguson, A. (2013). The heretic: Who is Thomas Nagel and why are so many of his fellow academics condemning him? *The Weekly Standard*, 18(27). Retrieved from http://www.weeklystandard.com/articles/heretic_707692.html

Ferrara, F., Pratt, D., & Robutti, O. (2006). The role and uses of technologies for the teaching of algebra and calculus. In A. Gutiérrez & P. Boero (Eds.), *Handbook of research on the psychology of mathematics education* (pp. 237–273). Rotterdam, The Netherlands: Sense Publishers.

Ferrara, F. & Nemirovsky, R. (2005). Connecting talk, gesture, and eye motion for the microanalysis of mathematics learning. In H. L. Chick, & J. L.

Vincent, (Eds.). Proceedings of the 29th Conference of the International Group for the Psychology of Mathematics Education, Vol. 1, pp. 138–142. Melbourne: PME.

Ferrara, F., & Savioli, K. (2009). Why could not a vertical line appear? Imagining to stop time. In M. Tzekaki, M. Kaldrimidou, & C. Sakonidis (Eds.), *Proceedings of the 33rd Conference of the International Group for the Psychology of Mathematics Education (PME 33)* (Vol. 3, pp. 33–40). Thessaloniki, Greece: PME.

Feser, E. (2010). *The last superstition: A refutation of the new atheism*. South Bend, IN: Saint Augustine's Press.

Fitzgerald, M. (2000) Is the cognitive style of the persons with the Asperger's Syndrome also a 'mathematical style'?, *Journal of Autism and Developmental Disorders*, *30*(2), 175–176.

(2002). Asperger's Disorder and mathematicians of genius. *Journal of Autism and Developmental Disorders*, *32*(1), 59–60.

(2004). *Autism and creativity: Is there a link between autism in men and exceptional ability?* London: Routledge.

Fomin, D., Genkin, S., & Itenberg, I. (1996). *Mathematical circles (Russian Experience)*. Providence, RI: American Mathematical Society.

Fuchs, L. S., Fuchs, D., Hamlett, C. L., & Appleton, A. C. (2002). Explicitly teaching for transfer: Effects on the mathematical problem-solving performance of students with mathematics disabilities. *Learning Disabilities Research & Practice*, *17*, 90–106.

Fuson, K. C. (1988). *Children's counting and concepts of number*. New York: Springer.

Gallese, V., & Lakoff, G. (2005). The brain's concepts: The role of the sensory-motor system in conceptual knowledge. *Cognitive Neuropsychology*, *21*, 1–25.

Gates, P. (2002). Issues of equity in mathematics education: Defining the problem, seeking solutions. In L. Haggarty (Ed.), *Teaching mathematics in secondary schools* (pp. 211–228). London: Routledge-Falmer.

Gee, J. P. (1991). A linguistic approach to narrative. *Journal of Narrative and Life History*, *1*, 15–39.

(2005). *An introduction to discourse analysis*. London & New York: Routledge.

Gifford, S. (2006). Dyscalculia: Myths and models. *Research in Mathematics Education*, *8*, 35–51.

Gleick, J. (1987). *Chaos: Making a new science*. London: Penguin Publishing.

Goguen, J. (1991). A categorical manifesto. *Mathematical Structures in Computer Science* *1*(1), 49–67.

Gough, N. (2004). RhizomANTically becoming-cyborg: Performing posthuman pedagogies. *Educational Philosophy and Theory*, *36*, 253–265.

Goodley, D. (2009). Bringing the psyche back into disability studies: The case of the body with/out organs. *Journal of Literary & Cultural Disability Studies*, 3(3), 257–272.

Grawemeyer, B. & Cox, R. (2008). The effects of users' background diagram knowledge and task characteristics upon information display selection. In G. Stapleton, J. Howse, & J. Lee (Eds.), *Diagrammatic Representation and Inference, 5th International Conference, Diagrams 2008, Proceedings* (pp. 321–334). Lecture Notes in Computer Science 5223. Berlin & Heidelberg: Springer.

Greiffenhagan, C. (in press). The materiality of mathematics: Presenting mathematics at the blackboard. *British Journal of Sociology*.

Grosz, E. (1994). *Volatile bodies: Toward a corporeal feminism*. Bloomington: Indiana University Press.

Grosz, E. A. (Ed.). (1999). *Becomings: Explorations in time, memory and futures*. Ithaca, NY: Cornell University Press.

Grosz, E. A. (2001). *Architecture from the outside: Essays on virtual and real space*. Cambridge, MA: MIT Press.

(2010). Political matters: Feminism, materialism, and freedom. In D. H. Coole & S. Frost (Eds.), *New materialisms: Ontology, agency, and politics* (pp. 139–157). Durham, NC: Duke University Press.

Gutstein, E. (2000). Increasing equity: Challenges and lessons from a state systemic initiative. In W. G. Secada (Ed.), *Changing the faces of mathematics: Perspectives on multiculturalism and gender equity* (pp. 25–36). Reston, VA: NCTM Publishing.

(2006). *Reading and writing the world with mathematics: Toward a pedagogy for social justice*. New York: Routledge.

Hacking, I. (2013). Why is there philosophy of mathematics at all? In M. Pitici & D. Mumford (Eds.), *The best writing on mathematics 2012* (pp. 234–254). Princeton: Princeton University Press.

Hadamard, J. (1945). *The psychology of invention in the mathematical field*. Princeton: Princeton University Press.

Hall, R., & Nemirovsky, R. (2011). Histories of modal engagements with mathematical concepts: A theory memo. Retrieved from http://www.sci.sdsu.edu/tlcm/all-articles/Histories_of_modal_engagement_with_mathematical_concepts.pdf

Hall, S. (1996). Introduction: Who needs identity? In S. Hall & P. de Guy (Eds.), *Questions of cultural identity* (pp. 1–17). London: Sage Publications.

Hallahan, D. P., & Mock, D. R. (2006). A brief history of the field of learning disabilities. In H. L. Swanson, K. R. Harris, & S. Graham (Eds.), *Handbook of learning disabilities* (pp. 16–29). New York: The Guilford Press.

Halliday, M. A. K. (1985). *An introduction to functional grammar.* Victoria, Australia: Edward Arnold.

(1991). Towards probabilistic interpretations. In E. Ventola (Ed.), *Trends in Linguistics Studies and Monographs: Vol. 55. Functional and systemic linguistics approaches and uses* (pp. 39–61). Berlin: Mouton de Gruyter.

(1993). Towards a language-based theory of learning. *Linguistics and Education*, *5*, 93–116.

(2007). The notion of 'context' in language education. In J. J. Webster (Ed.), *Language and education* (pp. 269–290). London: Continuum Press.

Hardy, G. H. (1929). Mathematical proof. *Mind, New Series*, *38*(149), 1–25.

(1940). *A mathematician's apology*. Cambridge: Cambridge University Press.

Harroway, D. (2008). *When species meet.* Minneapolis: University of Minnesota Press.

Healy, L., & Fernandes, S. H. A. A. (2011). The role of gestures in the mathematical practices of those who do not see with their eyes. *Educational Studies in Mathematics*, 77, 157–174.

Healy, L., & Powell, A. (2013). Understanding and overcoming 'disadvantage' in learning mathematics. In M. A. Clements, A. Bishop, C. Keitel-Kreidt, J.

Kilpatrick, & F. K.-S. Leung (Eds.), *Third international handbook of mathematics education* (pp. 69–100). Berlin: Springer.

Hehir, T. (2005). *New directions in special education: Eliminating ableism in policy and practice*. Cambridge, MA: Harvard Education Press.

Heidegger, M. (1977). *Sein und Zeit* [Being and time]. Tübingen, Germany: Max Niemeyer.

Henry, M. (2000). *Incarnation: Une philosophie de la chair* [Incarnation: A philosophy of the flesh]. Paris: Éditions du Seuil.

(2005). *Voir l'invisible: Sur Kandinsky* [Seeing the invisible: On Kandinsky]. Paris: Presses Universitaires de France.

Herbel-Eisenmann, B. (2007). From intended curriculum to written curriculum: Examining the 'voice' of a mathematics textbook. *Journal for Research in Mathematics Education*, *38*, 344–369.

Herbel-Eisenmann, B., & Wagner, D. (2010). Appraising lexical bundles in mathematics classroom discourse: Obligation and choice. *Educational Studies in Mathematics*, *75*(1), 43 – 63.

Herbst, P. (2002). Establishing a custom of proving in American school geometry: Evolution of the two-column proof in the early twentieth century. *Educational Studies in Mathematics*, *49*(3), 283–312.

Hersh, R. (1997). *What is mathematics, really?* New York: Oxford University Press.

Hersh, R., & John-Steiner, V. (2011). *Loving and hating mathematics: Challenging the myths of mathematical life*. Princeton: Princeton University Press.

Hewitt, D. (1999). Arbitrary and necessary: A way of viewing the mathematics curriculum. *For the Learning of Mathematics*, *19*(3), 2–9.

Hickey-Moody, A. (2009). Little war machines: Posthuman pedagogy and its media. *Journal of Literary & Cultural Disability Studies*, *3*(3), 273–280.

Hoffmann, M. H. G. (2005). Signs as means for discoveries: Peirce and his concepts of 'diagrammatic reasoning,' 'theorematic deduction,' 'hypostatic abstraction,' and 'theoric transformation'. In M. H. G. Hoffmann, J. Lenhard, & F. Seeger (Eds.), *Activity and sign: Grounding mathematics education* (pp. 45–56). New York: Springer.

Hooks, B. (2000). *Where we stand: Class matters*. New York: Routledge University.

Hughes, M. (1986). *Children and number: Difficulties in learning mathematics*. Oxford: Blackwell.

Husserl, E. (2001). *Analysis concerning passive and active synthesis: Lectures on transcendental logic*. Dordrecht, The Netherlands: Kluwer Academic Publishers.

Hutchins, E. (1995). *Cognition in the wild*. Cambridge, MA: MIT Press.

Hwang, S., & Roth, W.-M. (2011). *Scientific and mathematical bodies: The interface of culture and mind*. Rotterdam, The Netherlands: Sense Publishers.

Ilyenkov, E. V. (1977). *Dialectical logic*. Moscow: Progress Publishers.

Ingold, T. (2007). *Lines: A brief history*. Abingdon, Oxon: Routledge.

(2008). When ANT meets SPIDER: Social theory for arthropods. In C. Knappett & L. Malafouris (Eds.), *Material agency: Towards a non-anthropocentric approach* (pp. 209–215). New York: Springer.

(2010). Bringing things to life: Creative entanglements in a world of materials (NCRM Working Paper). University of Manchester: Realities/Morgan Centre.

(2011). *Being alive: Essays on movement, knowledge and description*. London: Routledge.

Ivins, W. (1938/1975). *On the rationalization of sight*. New York: Da Capo Press. (Original work published 1938)

Ivins, W. M., Jr., (1945/1964). *Art and geometry: A study of space intuitions*. New York: Dover Publications. (Original work published 1945)

Jackiw, N. (1991, 2001). *The Geometer's Sketchpad* Software. Emeryville, CA: Key Curriculum Press.

(2006). Mechanism and magic in the psychology of dynamic geometry. In N. Sinclair, D. Pimm, & W. Higginson (Eds.), *Mathematics and aesthetics: New approaches to an ancient affinity* (pp. 145–159). New York: Springer.

Jackiw, N., & Sinclair, N. (2009). Sounds and pictures: Dynamism and dualism in dynamic geometry. *ZDM: The International Journal on Mathematics Education, 41*, 413–426.

Jaffe, A., & Quinn, F. (1993). Theoretical mathematics: Toward a cultural synthesis of mathematics and theoretical physics. *Bulletin of the American Mathematical Society, 29*(1), 1–13.

Jaffe, A. & Quinn, F. (1994). Response to comments on 'theoretical mathematics'. *Bulletin of the American Mathematical Society, 30*(2), 208–211.

James, I. (2002). *Remarkable mathematicians: From Euler to von Neumann*. Cambridge: The Mathematical Association of America/Cambridge University Press.

(2003). Autism in mathematicians. *The Mathematical Intelligencer, 25*(4), 62–65.

(2004). *Remarkable physicists: From Galileo to Yukawa*. Cambridge: Cambridge University Press.

Jaworski, A., & Coupland, N. (Eds.). (1993). *The discourse reader*. New York: Routledge.

Jefferson, G. (2004). Glossary of transcript symbols with an introduction. In G. H. Lerner (Ed.), *Conversation analysis: Studies from the first generation* (pp. 13–23). Philadelphia: John Benjamins.

Jullien, C. (2008). *Esthétique et mathématiques: Une exploration goodmanienne*. Rennes, France: Presses Universitaires de Rennes.

Kanner, L. (1943). Autistic disturbances of affective contact. *Nervous Child, 2*, 217–250.

Kastberg, S., Norton, A., & Klerlein, J. (2009). Trusting students. *Mathematics Teaching in the Middle School, 17*, 423–429.

Kilpatrick, J., Swafford, J., & Findell, B. (Eds.). (2001). *Adding it up: Helping children learn mathematics*. Washington, DC: The National Academies Press.

Kirby, V. (2011). *Quantum anthropologies: Life at large*. Durham, NC: Duke University Press.

Kittler, F. (1999). *Gramophone, film, typewriter*. Stanford, CA: Stanford University Press.

Klibansky, R., Panofsky, E., & Saxl, F. (1964). *Saturn and melancholy: Studies in the history of natural philosophy, religion and art*. London: Nelson.

Kline, M. (1953). *Mathematics in Western culture*. London: George Allen and Unwin.

Knoespel, K. (2000). Introduction: Diagrammatic writing and the configuration of space. In G. Châtelet, *Figuring space* (pp. ix–xxiii). Dordrecht, The Netherlands: Kluwer Academic Publishers.

Krause, R., & Rölli, M. (2008). Micropolitical associations. In I. Buchanan & N. Thoburn (Eds.), *Deleuze and politics* (pp.240–254). Edinburgh: Edinburgh University Press.

Kristeva, J. (1980). *Desire in language: A semiotic approach to literature and art.* (T. Gora, A. Jardine, Trans. & L. S. Roudiez, Ed.). New York: Columbia Press.

Krull, W. (1930/1987). The aesthetic viewpoint in mathematics. *The Mathematical Intelligencer*, *9*(1), 48–52. (Original work published 1930)

Labov, W. (2006). Narrative pre-construction. *Narrative Inquiry*, *16*(1), 37–45.

Labov, W., & Waletzky, J. (1967). Narrative analysis. In J. Helm (Ed.), *Essays on the verbal and visual arts* (pp. 12–44). Seattle, WA: Washington University Press.

Lakatos, I. (1976). *Proofs and refutations: The logic of mathematical discovery*. J. Worrall & E. Zahar (Eds.). Cambridge: Cambridge University Press.

(1978). *Mathematics, science and epistemology: Philosophical papers* (Vol. 2). J. Worral, & G. Currie (Eds.). Cambridge: Cambridge University Press.

Lakoff, G., & Núñez, R. (2000). *Where mathematics comes from: How the embodied mind brings mathematics into being*. New York: Basic Books.

Lampert, M. (1990). When the problem is not the question and the solution is not the answer: Mathematical knowing and teaching. *American Educational Research Journal*, *27*(1), 29–64.

Lather, P. (1986). Issues of validity in openly ideological research: Between a rock and a soft place. *Interchange*, *17*(4), 63–84.

(2007). *Getting lost: Feminist efforts toward a double(d) science*. New York: SUNY Press.

Latour, B. (1990). Drawing things together. In M. Lynch & S. Woolgar (Eds.), *Representation in scientific practice* (pp. 19–68). Cambridge, MA: MIT Press.

– (1993). *We have never been modern*. Cambridge, MA: Harvard University Press.

(2005). *Reassembling the social: An introduction to actor-network-theory*. Oxford: Oxford University Press.

Lave, J., & Wenger, E. (1991). *Situated learning: Legitimate peripheral participation*. Cambridge: Cambridge University Press.

Lecercle, J-J. (2002). *Deleuze and language*. Basingstoke: Palgrave Macmillan.

Leikin, R. (2009). Multiple proof tasks: Teacher practice and teacher education. In F. Lin, F. Hsieh, G. Hanna, & M. de Villers (Eds.), *Proceedings of the ICMI Study-19 Conference: Proof and Proving in Mathematics Education* (Vol. 2, pp. 31–36). Taipei: ICMI.

Leikin, R., Berman, A., & Koichu, B. (Eds.). (2009). *Creativity in mathematics and the education of gifted students*. Rotterdam, The Netherlands: Sense Publishers.

Lemke, J. (2000). Opening up closure: Semiotics across scales. In J. Chandler & G. V. D. Vijver (Eds.), *Closure: Emergent organizations and their dynamics* (pp. 100–111). New York: New York Academy of Science Press.

(2002). Ideology, intertextuality and the communication of science. In P. H. Fries, M. Cummings, D. Lockwood, & W. Spruiell (Eds.), *Relations and functions within and around language* (pp. 32–55). New York: Continuum.

Leonard, J., & Jackson Dantley, S. (2005). Breaking through the ice: Dealing with issues of diversity in mathematics and science education courses. In A. Rodriguez & R. Kitchen (Eds.), *Preparing mathematics and science teachers for diverse classrooms: Promising strategies for transformative pedagogy* (pp. 87–118). Mahwah, NJ: Lawrence Erlbaum.

Leont'ev, A. N. (1978). *Activity, consciousness, and personality*. New Jersey: Prentice-Hall.

Lerman. S. (2005, May 27). *Identity in mathematics education.* Keynote presentation at The Canadian Mathematics Education Study Group, University of Ottawa.

Linton, S. (1998). *Claiming disability: Knowledge and identity.* New York: New York University Press.

Lockhart, P. (2009). *A mathematician's lament: How school cheats us out of our most fascinating and imaginative art form.* New York: Bellevue Literary Press.

Loveland, K. (1991). Social affordances and interaction: Autism and the affordances of the human environment. *Ecological Psychology, 3*(2), 99–119.

Lundin, S. (2011). Hating school, loving mathematics: On the ideological function of critique and reform in mathematics education. *Educational Studies in Mathematics, 80*(1), 73–85.

MacLure, M. (2010, May 4). *Facing Deleuze: Affect in education and research.* Presentation at The American Educational Research Association, Denver, CO.

Malafouris, L. (2008). Between drains, bodies and things: *Tectonoetic* awareness and extended self. *Philosophical Transactions of the Royal Society B, 363,* 1993–2002.

Malloy, C. (2004). Equity in mathematics education is about access. In R. N. Rubenstein & G. W. Bright (Eds.), *Perspectives on the teaching of mathematics* (pp. 1–14). Reston, VA: NCTM Publishing.

Manning, E. (2009). *Relationscapes: Movement, art, philosophy.* Cambridge, MA: MIT Press.

Manning, E., & Massumi, B. (2010, November 11). Coming alive in a world of texture: For neurodiversity. Keynote presentation at the Thinking – Resisting – Reading the Political Conference, Giessen, Germany. Retrieved from http://www.youtube.com/watch?v=DqUaEcO30T0

Marion, J.-L. (2002). *Being given: Toward a phenomenology of givenness.* Stanford, CA: Stanford University Press.

(2004). *The crossing of the visible.* Stanford, CA: Stanford University Press.

Marschark, M., Spencer, P. E., Adams, J., & Sapere, P. (2011). Evidence-based practice in educating deaf and hard-of-hearing children: Teaching to their cognitive strengths and needs. *European Journal of Special Needs Education, 26,* 3–16.

Marsh, M. (2002). Examining the discourses that shape our teacher identities. *Curriculum Inquiry, 32*(4), 453–469.

Martin, J. R. (1993). Genre and literacy: Modeling context in educational linguistics. *Annual Review of Applied Linguistics, 13,* 141–172.

(2007). Construing knowledge: A functional linguistic perspective. In F. Christie & J. R. Martin (Eds.), *Language, knowledge and pedagogy: Linguistic and sociological perspectives* (pp. 34–64). London: Continuum International Publishing Group.

Massumi, B. (2002). *Parables for the virtual: Movement, affect, sensation.* Durham, NC: Duke University Press.

Maturana, H. R., & Varela, F. J. (1987). *The tree of knowledge: The biological roots of human understanding.* Boston: Shambhala.

McCrink, K., & Spelke, E. S. (2010). Core multiplication in childhood. *Cognition, 116*(2), 204–216.

McCullough, W. S. (1965). *Embodiments of mind.* Cambridge, MA: MIT Press.

McNeill, D. (2001). Analogic/analytic representations and cross-linguistic differences in thinking for speaking. *Cognitive Linguistics, 11,* 43–60.

(2003). Pointing and morality in Chicago. In S. Kita (Ed.), *Pointing: Where language, culture and cognition meet* (pp. 139–170). Mahwah, NJ: Erlbaum.

(2005). *Gesture and thought*. Chicago: University of Chicago Press.

Mendick, H. (2006). *Masculinities in mathematics*. Maidenhead, UK: Open University Press.

Merleau-Ponty, M. (1945). *Phenomenologie de la perception* [Phenomenology of perception]. Paris: Gallimard.

(1962/2004). Excerpt from *Phenomenology of perception*. (K. Paul, Trans.). In T. Baldwin (Ed.), *Basic Writings*. New York: Routledge.

Miller, J. (1992). Teachers, autobiography, and curriculum: Critical and feminist perspectives. In B. B. Swadner & S. Kessler (Eds.), *Reconceptualizing the early childhood curriculum* (pp. 103–122). New York: Teachers College Press.

Mitchell, W. J. T. (2001). Seeing disability [Special issue]. *Public Culture*, *13*(3), 391–397.

Molland, G. (1991/1995). Implicit versus explicit geometrical methodologies: The case of construction. In G. Molland, *Mathematics and the medieval ancestry of physics* (pp. 181–196). Brookfield, VT: Variorum. (Original work published 1991)

Morgan, C. (2005). Words, definitions and concepts in discourses of mathematics teaching and learning. *Language and Education*, *19*(2), 103–117.

(2006). What does social semiotics have to offer mathematics education research? *Educational Studies in Mathematics*, *61*, 219–245.

Moses, R. P., & Cobb, C. E. (2001). *Radical equations: Civil rights from the Mississippi to the Algebra Project*. Boston: Beacon Press.

Movshovitz-Hadar, N. (1988). School mathematics theorems: An endless source of surprise. *For the Learning of Mathematics*, *8*(3), 34–39.

Mukhopadhyay, S., & Greer, B. (2001). Modeling with purpose: Mathematics as a critical tool. In B. Atweh, H. Forgasz, & B. Nebres (Eds.), *Socio-cultural research on mathematics education: An international perspective* (pp. 295–312). Mahwah, NJ: Lawrence Erlbaum.

Munn, P., & Reason, R. (2007). Arithmetic difficulties: Developmental and instructional perspectives. *Educational and Child Psychology*, *24*(2), 5–14.

Nagel, T. (2012). *Mind and cosmos: Why the materialist neo-Darwinian conception of nature is almost certainly false*. New York: Oxford University Press.

Nancy, J.-L. (2007). *Listening*. New York: Fordham University Press.

Nemirovsky, R. (2003). Three conjectures concerning the relationship between body activity and understanding mathematics. In N. A. Pateman, B. J. Dougherty, & J. T. Zilliox (Eds.), *Proceedings of the 27th Conference of the International Group for the Psychology of Mathematics Education (PME 27)* (Vol. 1, pp. 103–135). Honolulu: PME.

Nemirovsky, R., & Ferrara, F. (2009). Mathematical imagination and embodied cognition. *Educational Studies in Mathematics*, *70*, 159–174.

Nemirovsky, R., Rasmussen, C., Sweeney, G., & Wawro, M. (2012). When the classroom floor becomes the complex plane: Addition and multiplication as ways of bodily navigation. *Journal of the Learning Sciences*, *21*(2), 287–323.

Netz, R. (1999). *The shaping of deduction in Greek mathematics: A study in cognitive history*. Cambridge: Cambridge University Press.

(2009). *Ludic proof: Greek mathematics and the Alexandrian aesthetic*. New York: Cambridge University Press.

Netz, R., Noel, W., Wilson, N., & Tchernetska, N. (2011). *The Archimedes palimpsest* (Vols. 1–2). Cambridge: Cambridge University Press.

Niremberg, D., & Niremberg, R. (2011). Badiou's number: A critique of mathematical ontology. *Critical Inquiry*, *37*(4), 583–614.

Novick, L. R. (2004). Diagram literacy in preservice math teachers, computer science majors, and typical undergraduates: The case of matrices, networks, and hierarchies. *Mathematical Thinking and Learning*, *6*, 307–342.

Nunes, T. (2004). *Teaching mathematics to deaf children*. London: Whurr Publishers.

Núñez, R. (2006). Do real numbers really move? Language, thought, and gesture: The embodied cognitive foundations of mathematics. In R. Hersh (Ed.), *18 unconventional essays on the nature of mathematics* (pp. 160–181). New York: Springer.

Núñez, R., Edwards, L., & Matos, J. F. (1998). Embodied cognition as grounding for situatedness and context in mathematics education. *Educational Studies in Mathematics*, *39*, 45–65.

Nunokawa, K. (2001). Surprises in mathematics lessons. *For the Learning of Mathematics*, *21*(3), 43–50.

(2004). Solvers' making of drawings in mathematical problem solving and their understanding of the problem situations. *International Journal of Mathematical Education in Science and Technology*, *35*, 173–183.

(2006). Using drawings and generating information in mathematical problem solving processes. *Journal of Mathematics, Science and Technology Education*, *2*(3), 33–54.

O'Halloran, K. L. (2005). *Mathematical discourse: Language, symbolism and visual images*. London & New York: Continuum.

(2010). The semantic hyperspace: Accumulating mathematical knowledge across semiotic resources and modes. In F. Christie & K. Maton (Eds.), *Disciplinarity: Functional linguistic and sociological perspectives* (pp. 217–236). London & New York: Continuum.

O'Halloran, K. L., & Smith, B. A. (2011). Multimodal studies. In K. L. O'Halloran & B. A. Smith. (Eds.), *Multimodal studies: Exploring issues and domains* (pp. 1–13). London & New York: Routledge.

Ober, J. (2007). *The original meaning of 'democracy': Capacity to do things, not majority rule* (Princeton/Stanford Working Papers in Classics). Retrieved from http://ssrn.com/abstract=1024775

Olive, J., & Steffe, L. (2002). Schemes, schemas and director systems (An integration of Piagetian scheme theory with Skemp's model of intelligent learning). In D. O. Tall & M. O. J. Thomas (Eds.), *Intelligence, learning and understanding in mathematics: A tribute to Richard Skemp* (pp. 97–130). Flaxton, QLD: Post Pressed.

Ong, W. J. (1982). *Orality and literacy: The technologizing of the word*. London: Methuen.

Otte, M. (2005). Mathematics, sign and activity. In M. H. G. Hoffman, J. Lenhard, & F. Seeger (Eds.), *Activity and sign: Grounding mathematics education* (pp. 9–22). New York: Springer.

Overboe, J. (2001). A critique of the ableist model of disability as lack. Paper presented at the Society for Literature and Science Conference 2001: Technologies, Bodies, Narratives: The Accountability of Scientific and Medical Practices Session, Buffalo, NY.

(2007a). Ableist limits on self-narration: The concept of post-personhood. In V. Raoul, C. Canam, A. Henderson, & C. Paterson (Eds.), *Unfitting stories: Narrative approaches to disease, disability, and trauma* (pp. 175–182). Waterloo: Wilfrid Laurier University Press.

(2007b). Disability and genetics: Affirming the bare life (the state of exception). In S. J. Reuters & K. Neves-Graca (Eds.), Genes and society: Looking back on the future [Special issue]. *Canadian Review of Sociology and Anthropology*, *44*(2), 219–235.

(2007c). Vitalism: Subjectivity exceeding racism, sexism, and (psychiatric) ableism. In P. Parekh (Ed.), Intersecting gender and disability Perspectives in rethinking postcolonial identities [Special issue]. *Wagadu, Journal of Transnational Women's and Gender Studies*, *4*, 22–34.

(2009a). Affirming an impersonal life: A different register for disability studies. *Journal of Literary & Cultural Disability Studies*, *3*(3), 241–256.

(2009b). Introduction: Deleuze, disability, and difference. *Journal of Literary & Cultural Disability Studies*, *3*(3), 217–220.

Papert, S. (1980). *Mindstorms: Children, computers and powerful ideas*. New York: Basic Books.

Parkinson, G. H. R., & Morris, M. (Eds. & Trans.). (1973). *Leibniz: Philosophical writings*. London: Everyman's Library.

Paterson, K., & Hughes, B. (1999). Disability studies and phenomenology: The carnal politics of everyday life. *Disability and Society*, *14*(5), 597–610.

Pickering, A. (1995). *The mangle of practice: Time, agency and science*. Chicago: University of Chicago Press.

Pimm, D. (1993). From should to could: Reflections on possibilities of mathematics teacher education. *For the Learning of Mathematics*, *13*(2), 27–32.

(2006). Drawing on the image in mathematics and art. In N. Sinclair, D. Pimm, & W. Higginson (Eds.), *Mathematics and the aesthetic: New approaches to an ancient affinity* (pp. 160–189). New York: Springer.

Pimm. D., Beisiegel, M., & Meglis, I. (2008). Would the real Lakatos please stand up? *Interchange: A Quarterly Review of Education*, *39*(4), 469–481.

Pimm, D., & Sinclair, N. (2009). Audience, style and criticism. *For the Learning of Mathematics*, *29*(2), 23–27.

Plucker, J. A., & Beghetto, R. A. (2004). Why creativity is domain general, why it looks domain specific, and why the distinction does not matter. In R. J. Sternberg, E. L. Grigorenko, & J. L. Singer (Eds.), *Creativity: From potential to realization* (pp. 153–168). Washington, DC: American Psychological Association.

Poincaré, H. (1905). *Science and hypothesis*. New York: The Science Press.

(1908/1956). Mathematical creation. In J. Newman (Ed.), *The world of mathematics* (Vol. 4, pp. 2041–2050). New York: Simon and Schuster. (Original work published 1908)

Pólya, G. (1981). *Mathematical discovery: On understanding, learning, and teaching problem solving* (Vol. 2). New York: John Wiley and Sons.

Popkewitz, T. (1998). *Struggling for the soul: The politics of schooling and the construction of the teacher.* New York: Teachers College Press.

(2004). The alchemy of the mathematics curriculum: Inscriptions and the fabrication of the child. *American Educational Research Journal, 41*(1), 3–34.

Povey, H., Burton, L., Angier, C., & Boylan, M. (2004). Learners as authors in the mathematics classroom. In B. Allen & S. Johnston-Wilder (Eds.), *Mathematics education: Exploring the culture of learning* (pp. 43–56). New York: Routledge-Falmer.

Radden, J. (2000). *The nature of melancholy: From Aristotle to Kristeva.* New York: Oxford University Press.

Radford, L. (2003). Gestures, speech, and the sprouting of signs: A semiotic-cultural approach to students' types of generalization. *Mathematical Thinking and Learning, 5*(1), 37–70.

(2004). Rescuing perception: Diagrams in Peirce's theory of cognitive activity. Paper presented at ICME 10, Denmark, Copenhagen.

(2008). The ethics of being and knowing: Towards a cultural theory of learning. In L. Radord, G. Schubring, & F. Seeger (Eds.), *Semiotics in mathematics education: Epistemology, history, classroom and culture* (pp. 215–234). Rotterdam, The Netherlands: Sense Publishers.

(2009). 'No! He starts walking backwards!': Interpreting motion graphs and the question of space, place and distance. *ZDM: The International Journal on Mathematics Education, 41*, 467–480.

Radford, L., & Roth, W.-M.(2011). *A cultural-historical perspective on mathematics teaching and learning.* Rotterdam, The Netherlands: Sense Publishers.

Rajchman, J. (2000). *The Deleuze connections.* Cambridge, MA: MIT Press.

Rancière, J. (2002). The aesthetic revolution and its outcomes: Emplotments of autonomy and heteronomy. *New Left Review, 14*(March/April), 133–151.

(2004). *The politics of aesthetics: The distribution of the sensible.* (G. Rockhill, Trans.). New York: Continuum.

(2009). Contemporary art and the politics of aesthetics. In B. Hinderliter, W. Kaizen, & V. Maimon, *Communities of sense: Rethinking aesthetics and politics* (pp. 31–50). Durham, NC: Duke University Press.

(2010). *Dissensus: On politics and aesthetics.* (S. Corcoran, Trans. & Ed.). London: Continuum Press.

Raoul, V., Canam, C., Henderson, A., & Paterson, C. (2007). Making sense of disease, disability, and trauma: Normative and disruptive stories. In V. Raoul, C. Canam, A. Henderson, & C. Paterson (Eds.), *Unfitting stories: Narrative approaches to disease, disability, and trauma* (pp. 3–10). Waterloo: Wilfrid Laurier University Press.

Redman, P. (2005). The narrative formation of identity revisited: Narrative construction, agency and the unconscious. *Narrative Inquiry, 15*(1), 25–44.

Rée, J. (1999). *I see a voice: A philosophical history of language, deafness and the senses.* London: Harper Collins.

Rice, S. (2002). The social construction of 'disabilities': The role of law. *Educational Studies, 33*, 169–180.

Riskin, J. (2002). *Science in the age of sensibility: The sentimental empiricists of the French Enlightenment.* Chicago: University of Chicago Press.

Ringrose, J. (2010). Beyond discourse: Using Deleuze and Guattari's schizoanalysis to explore affective assemblages, heterosexually striated space, and lines of flight online and at school. *Educational Philosophy and Theory*, *43*(6), 1–21.

Robutti, O. (2006). Motion, technology, gesture in interpreting graphs. *The International Journal for Technology in Mathematics Education*, *13*(30), 117–126.

Rodriguez, A. J., & Kitchen, R. S. (Eds.). (2005). *Preparing mathematics and science teachers for diverse classrooms: Promising strategies for transformative pedagogy*. Mahwah, NJ: Lawrence Erlbaum.

Rogers, M. P., Malancharuvil-Berkes, E., Mosley, M., Hui, D., & O'Garro Joseph, G. (2005). Critical discourse analysis in education: A review of the literature. *Review of Educational Research*, *75*(3), 365–416.

Rota, G. (1997). *Indiscrete thoughts*. Boston: Birkhäuser.

Roth, W.-M. (2010). Incarnation: Radicalizing the embodiment of mathematics. *For the Learning of Mathematics*, *30*(2), 8–17.

(2011). *Geometry as objective science in elementary classrooms: Mathematics in the flesh*. New York: Routledge.

Rotman, B. (2000). *Mathematics as sign: Writing, imagining, counting*. Stanford: Stanford University Press.

(2008). *Becoming beside ourselves: The alphabet, ghosts, and distributed human beings*. Durham, NC: Duke University Press.

(2012). Topology, algebra, diagrams. *Theory, Culture, Society*, *29*, 247–260.

Rowland, T. (1995). Hedges in mathematics talk: Linguistic pointers to uncertainty. *Educational Studies in Mathematics*, *29*(4), 327–353.

(2000). *The pragmatics of mathematics education: Vagueness in mathematical discourse*. London: Falmer Press.

Russell, B. (1903). *The principles of mathematics*. Cambridge: Cambridge University Press.

(1919). The study of mathematics. In *Mysticism and logic: And other essays* (p. 60). Cambridge: Longmans, Green & Company.

Sacks, O. (1984). *A leg to stand on*. New York: Touchstone.

Sasaki, B. (2002). Toward a pedagogy of coalition. In A. A. Macdonald & S. Sanchez-Casal (Eds.), *Twenty-first-century feminist classrooms: Pedagogies of identity and difference* (pp. 31–58). New York: Palgrave Macmillan.

Sawyer, W. (1955). *Prelude to mathematics*. Harmondsworth: Pelican.

Saxe, G. B., Shaughnessy, M. M., Shannon, A., Langer-Osuna, J. M., Chinn, R., & Gearhart, M. (2007). Learning about fractions as points on the number line. In *The learning of mathematics, 69th yearbook of The National Council of Teachers of Mathematics* (pp. 221–236). Reston, VA: NCTM Publishing.

Schegloff, E. A. (1997). Whose text? Whose context? *Discourse and society*, *8*(2), 165–87.

Scher, D. (1995). *Exploring conic sections with The Geometer's Sketchpad*. Berkeley, CA: Key Curriculum Press.

Schmittau, J. (2003). Cultural historical theory and mathematics education. In A. Kozulin, B. Gindis, S. Miller, & V. Ageyev (Eds.), *Vygotsky's educational theory in cultural context* (pp. 225–245). Cambridge: Cambridge University Press.

Schrödinger, E. (1935/1983). The present situation in quantum mechanics. In J. A. Wheeler & W. Zurek (Eds.), *Quantum theory and measurement* (pp.152–167). Princeton: Princeton University Press. (Original work published 1935)

Schulze, A., & Sevenoak, F. (1913). *Plane geometry*. New York: Macmillan.

Semetsky, I. (2006). *Deleuze, education, and becoming*. Rotterdam, The Netherlands: Sense Publishers.

Serres, M. (2008). *The five senses: A philosophy of mingled bodies*. (M. Sankey & P. Cowley, Trans.). London: Continuum International Publishing Group.

(2011). *Variations on the body*. (R. Burke, Trans.). Minneapolis: University of Minnesota Press.

Sfard, A. (2008). *Thinking as communicating: Human development, the growth of discourses, and mathematizing*. Cambridge: Cambridge University Press.

Shaffer, D. W., & Kaput, J. (1999). Mathematics and virtual culture: A cognitive evolutionary perspective on technology and mathematics education. *Educational Studies in Mathematics*, *37*(2), 97–119.

Shakespeare, T., & Watson, N. (2001). The social model of disability: An outdated ideology? In S. N. Barnartt & B. M. Altman (Eds.), *Exploring Theories and Expanding Methodologies: Where We Are and Where We Need to Go Series: Research in social science and disability* (Vol. 2, pp. 9–28). Amsterdam & New York: JAI.

Shapiro, S. (2000). *Thinking about mathematics: The philosophy of mathematics*. New York: Oxford University Press.

Sheets-Johnstone, M. (2009). Animation: The fundamental, essential, and properly descriptive concept. *Continental Philosophy Review*, *42*, 375–400.

(2012). Movement and mirror neurons: A challenging and choice conversation. *Phenomenology and the Cognitive Sciences*, *11*(3), 385–401.

Simondon, G. (2005). *L'individuation à la lumière des notions de forme et d'information*. Grenoble: Editions Jérôme Millon.

Sinclair, N. (2010). Knowing more than we can tell. In B. Sriraman & L. English (Eds.). *Theories of mathematics education: Seeking new frontiers* (pp. 591–608). Berlin: Springer.

Sinclair, N., de Freitas, E., & Ferrara, F. (2013). Virtual encounters: The murky and furtive world of mathematical inventiveness. *ZDM : The International Journal on Mathematics Education*, *45*(2), 239–252.

Sinclair, N., & Gol Tabaghi, S. (2010). Drawing space: Mathematicians' kinetic conceptions of eigenvectors. *Education Studies in Mathematics*, *74*(3), 223–240.

Sinclair, N., & Jackiw, N. (2011). On the origins of dynamic number in the breakdown of structural, metaphoric, and historic conceptions of human mathematics. In P. Liljedahl, S. Oesterle, & D. Allan (Eds.), *Proceedings/Actes 2010 Annual Meeting/Rencontre Annuelle* (pp. 137–146). Burnaby, BC: CMESG/GCEDM.

Sinclair, N., & Pimm, D. (2009). The many and the few: Mathematics, democracy and the aesthetic. *Educational Insights*, *13*(1). Retrieved from http://educationalinsights.ca/v13n01/articles/sinclair_pimm/index.html

Skemp, R. (1979). *Intelligence, learning and action – A foundation for theory and practice in education*. Chichester, UK: Wiley.

Skovsmose, O. (1994). *Towards a philosophy of critical mathematics education*. London: Springer.

Skovsmose, O., & Borba, M. (2004). Research methodology and critical mathematics education. In P. Valero & R. Zevenbergen (Eds.), *Researching the socio-political dimensions of mathematics education: Issues of power in theory and methodology* (pp. 207–226). New York: Kluwer Academic Publishers.

Skovsmose, O., & Valero, P. (2002). Democratic access to powerful mathematical ideas. In L. D. English (Ed.), *The handbook of international research in mathematics education* (pp. 383–408). Mahwah, NJ: Lawrence Erlbaum.

Smith, D. (2003). Deleuze on Bacon: Three conceptual trajectories in *The logic of sensation*. In G. Deleuze, *Francis Bacon: The logic of sensation* (pp. vii–xxvii). Minneapolis: University of Minnesota Press.

(2005). Deleuze on Leibniz: Difference, continuity, and the calculus. In S. H. Daniel (Ed.), *Current continental theory and modern philosophy* (pp. 127–147). Evanston, IL: Northwestern University Press. 127–147.

Smith, D. W. (2006). Axiomatics and problematics as two modes of formalization: Deleuze's epistemology of mathematics. In S. Duffy (Ed.), *Virtual mathematics: The logic of difference* (pp. 145–168). Manchester, UK: Clinamen Press.

(2007). The conditions of the new. *Deleuze Studies*, *1*(1), 1–21.

Snyder, S. L., & Mitchell, D. T. (2001). Re-engaging the body: Disability studies and the resistance to embodiment [Special issue]. *Public Culture*, *13*(3), 367–390.

Solomon, Y. (2009). *Mathematical literacy: Developing identities of inclusion*. New York: Routledge.

Staats, S. K. (2008). Poetic lines in mathematics discourse: A method from linguistic anthropology. *For the Learning of Mathematics*, *28*(2), 26–32.

Staats, S. K., & Batteen, C. (2010). Linguistic indexicality in algebra discourse. *The Journal of Mathematical Behavior*, *29*, 41–56.

Stanley, D. (2002). A response to Nunokawa's article: 'Surprises in mathematics lessons'. *For the Learning of Mathematics*, *22*(1), 15–16.

Stevens, R. (2000). Who counts what as math? Emergent and assigned mathematics problems in a problems-based classroom. In J. Boaler, (Ed.), *Multiple perspectives on mathematics teaching and learning* (pp. 105–145). London: Ablex Publishing.

(2012). The missing bodies of mathematical thinking and learning have been found. *Journal of the Learning Sciences*, *21*(2), 337–346.

Stockero, S. L., & van Zoest, L. R. (2011). Making student thinking public. *Mathematics Teacher*, *104*(9), 704–709.

Stokoe, E. H. (2005). Analysing gender and language. *Journal of sociolinguistics*, *9*(1), 118–33.

Stokoe, E., & Edwards, D. (2006). Story formation in talk-in-interaction. *Narrative Inquiry*, *16*(1), 56–65.

Straus, E. (1935/1963). *The primary world of the senses: A vindication of sensory experience*. (J. Needleman, Trans.). (2nd ed.). New York: Free Press (Original work published as *Vom Sinn der Sinne*, 1935)

Stylianou, D. A., & Silver, E. A. (2004). The role of visual representations in advanced mathematical problem solving: An examination of expert-novice similarities and differences. *Mathematical Thinking and Learning*, *6*(4), 353–387.

Tahta, D. (1980). About geometry. *For the Learning of Mathematics*, *1*(1), 2–9.

Tall, D. (2011). Crystalline concepts in long-term mathematical invention and discovery. *For the Learning of Mathematics*, *31*(1), 3–8.

Tannen, D. (2007). *Talking voices: Repetition, dialogue, and imagery in conversational discourse*. New York: Cambridge University Press.

Tate, W. F. (2005). Race, retrenchment, and the reform of school mathematics. In E. Gutstein & B. Peterson (Eds.), *Rethinking mathematics: Teaching social justice by the numbers* (pp. 31–40). Milwaukee, WI: Rethinking Schools.

Tate, W., & Rousseau, C. (2002). Access and opportunity: The political and social context of mathematics education. In L. D. English (Ed.), *Handbook of international research in mathematics education* (pp. 271–300). Mahwah, NJ: Lawrence Erlbaum.

Thibault, P. J. (2004a). *Agency and consciousness in discourse: Self-other dynamics as a complex system*. London & New York: Continuum.

(2004b). *Brain, mind, and the signifying body: An ecosocial semiotic theory*. London & New York: Continuum.

Titchkosky, T. (2006). *Disability, self, and society*. Toronto: University of Toronto Press.

(2007). *Reading and writing disability differently: The textured life of embodiment*. Toronto: University of Toronto Press.

Torrance, E. P. (1974). *Torrance tests of creative thinking*. Bensenville, IL: Scholastic Testing Service.

Tracy, K. (2002). *Everyday talk: Building and reflecting identities*. New York: Guilford Press.

UPIAS. (1976/1997). Fundamental Principles of Disability. Retrieved from http://www.leeds.ac.uk/disability-studies/archiveuk/UPIAS/fundamental%20principles.pdf

Valero, P. (2004). Socio-political perspectives on mathematics education. In P. Valero & R. Zevenbergen (Eds.), *Researching the socio-political dimensions of mathematics education: Issues of power in theory and methodology* (pp. 5–23). New York: Kluwer Academic Publishers.

Valero, P., & Zevenbergen, R. (Eds.). (2004). *Researching the socio-political dimensions of mathematics education: Issues of power in theory and methodology*. New York: Kluwer Academic Publishers.

Veel, R. (1999). Language, knowledge, and authority in school mathematics. In F. Christie (Ed.), *Pedagogy and the shaping of consciousness* (pp. 185–216). London: Cassell.

Vygotsky, L. (1934/1986). *Thought and language*. Cambridge, MA: MIT Press. (Original work published 1934)

Vygotsky, L., & Luria, A. (1994). Tool and symbol in child development. In R. van der Veer & J. Valsiner (Eds.), *The Vygotsky reader* (pp. 99–174). Oxford: Blackwell Publishers.

Walkerdine, V. (1988). *The mastery of reason: Cognitive development and the production of rationality*. London: Routledge.

(1998). *Counting girls out*. London: Falmer.

(1990). Difference, cognition, and mathematics education. *For the Learning of Mathematics, 10*(3), 51–56.

Walkington, J. (2005). Becoming a teacher: Encouraging development of teacher identity through reflective practice. *Asia-Pacific Journal of Teacher Education, 33*(1), 53–64.

Walshaw, M. (2004a). A powerful theory of active engagement. *For the Learning of Mathematics, 24*(3), 4–10.

(Ed.). (2004b). *Mathematics education within the postmodern*. Greenwich, CT: Information Age Publishing.

Walshaw, M., & Brown, T. (2012). Affective productions of mathematical experience. *Educational Studies in Mathematics*, *80*(1–2), 185–199.

Webb, T. (2008). Remapping power in educational micropolitics. *Critical Studies in Education*, *49*(2), 127–142.

Weil, A. (1992). *The apprenticeship of a mathematician*. (J. Gage, Trans.). Berlin: Birkhäuser.

Weissglass, J. (2000). No compromise on equity in mathematics education: Developing an infrastructure. In W. G. Secada (Ed.), *Changing the faces of mathematics: Perspectives on multiculturalism and gender equity* (pp. 5–24). Reston, VA; NCTM Publishing.

Wendell, S. (1993). Feminism, disability, and transcendence of the body. *Canadian Women's Studies*, *17*(4), 116–122.

West, T. G. (2004). *Thinking like Einstein: Returning to our virtual roots with the emerging revolution in computer information visualization*. Amherst, MA: Prometheus Books.

Wetherall, M. (1998). Positioning and interpretative repertoires: Conversation analysis and post-structuralism in dialogue. *Discourse and Society*, *9*(3). 431–456.

Whitehead, A. N. (1951). Mathematics and the good. In P. Schipp (Ed.). *The philosophy of Alfred North Whitehead* (pp. 666–681). La Salle: Open Court.

Whitin, P., & Whitin, D. J. (2002). Promoting communication in the mathematics classroom. *Teaching Children Mathematics*, *9*(4), 205–211.

Widdicombe, S. (1998). Identity as an analyst's and a participant's resource. In C. Antaki & S. Widdicombe (Eds.), *Identities in talk*. London: Sage.

Widdowson, H. G. (2004). *Text, context, pretext*. Malden, MA: Blackwell Publishing.

Wilensky, U. (1991). Abstract meditations on the concrete and concrete implications for mathematics education. In I. Harel & S. Papert (Eds.), *Constructionism* (pp. 193–204). Norwood, NJ: Ablex Publishing.

Williams, E., & Costall, A. (2000). Taking things more seriously: Psychological theories of autism and the material-social divide. In P. Graves (Ed.), *Matter, materiality and modern culture* (pp. 97–111). Routledge: London.

Wing, L. (1969). The handicaps of autistic children: A comparative study. *Journal of Child Psychology and Psychiatry*, *10*, 1–40.

Winks, J. (1999). *Critical pedagogy: Notes from the real world* (2nd ed.). New York: Allyn and Bacon.

Wittgenstein, L. (1958). *Philosophical investigations* (3rd ed.). New York: Macmillan.

(1978). *Remarks on the foundations of mathematics*. Oxford: Basil Blackwell.

Woffitt, R. (2005). *Conversation analysis and discourse analysis: A comparative and critical introduction*. Thousand Oaks, CA: Sage Publications.

Woodward, J. (2004). Mathematics education in the United States: Past to present. *Journal of Learning Disabilities*, *37*, 16–31.

Woodward, J., & Montague, M. (2002). Meeting the challenge of mathematics reform for students with LD. *Journal of Special Education*, *36*(2), 89–101.

Wortham, S. (2006). *Learning identity: The joint emergence of social identification and academic learning*. New York: Cambridge University Press.

Yackel, E., & Cobb, P. (1996). Sociomathematical norms, argumentation, and autonomy in mathematics. *Journal for Research in Mathematics Education*, 27(4), 458–477.

Yackel, E., & Hanna, G. (2003). Reasoning and proof. In J. Kilpatrick, W. G. Martin, & D. E. Schifter (Eds.), *A research companion to principles and standards for school mathematics* (pp. 227–236). Reston, VA: NCTM Publishing.

Yates, F. (1979). *The occult philosophy in the Elizabethan Age*. London: Routledge and Kegan Paul.

Young, K. (1999). Narrative embodiments: Enclaves of the self in the realm of medicine. In A. Jaworski & N. Coupland (Eds.), *The discourse reader* (pp. 428–441). New York: Routledge.

Zevenbergen, R. (2003). Teachers' beliefs about teaching mathematics to students from socially disadvantaged backgrounds: Implications for social justice. In L. Burton (Ed.), *Which way social justice and mathematics education?* (pp. 133–152). London: Praeger.

Author Index

Subject Index

BOOKS IN THE SERIES *(continued from page iii)*

Mind and Social Practice: Selected Writings of Sylvia Scribner
Ethel Tobach et al., Editors

Computation and Human Experience
Philip E. Agre

Situated Cognition: On Human Knowledge and Computer Representations
William J. Clancey

Communities of Practice: Learning, Meaning, and Identity
Etienne Wenger

Learning in Likely Places: Varieties of Apprenticeship in Japan
John Singleton, Editor

Talking Mathematics in School: Studies of Teaching and Learning
Magdalene Lampert and Merrie L. Blunk, Editors

Perspectives on Activity Theory
Yrjö Engeström, Reijo Miettinen, and Raija-Leena Punamäki, Editors

Dialogic Inquiry: Towards a Socio-cultural Practice and Theory of Education
Gordon Wells

Vygotskian Perspectives on Literacy Research: Constructing Meaning through Collaborative Inquiry
Carol D. Lee and Peter Smagorinsky, Editors

Technology in Action
Christian Heath and Paul Luff

Changing Classes: School Reform and the New Economy
Martin Packer

Building Virtual Communities: Learning and Change in Cyberspace
K. Ann Renninger and Wesley Shumar, Editors

Adult Learning and Technology in Working-Class Life
Peter Sawchuk

Vygotsky's Educational Theory in Cultural Context
Alex Kozulin, Boris Gindis, Vladimir S. Ageyev, and Suzanne M. Miller, Editors

Designing for Virtual Communities in the Service of Learning
Sasha A. Barab, Rob Kling, and James H. Gray, Editors

Bakhtinian Perspectives on Language, Literacy, and Learning
Arnetha F. Ball and Sarah Warshauer Freedman, Editors

Beyond Communities of Practice: Language, Power and Social Context
David Barton and Karin Tusting, Editors

Human-Machine Reconfigurations: Plans and Situated Actions, 2nd Edition
Lucy Suchman

Psychotherapy in Everyday Life
Ole Dreier

Thinking as Communicating: Human Development, the Growth of Discourses, and Mathematizing
Anna Sfard

Assessment, Equity, and Opportunity to Learn
Pamela A. Moss et al., Editors

From Teams to Knots: Studies of Collaboration and Learning at Work
Yrjö Engeström

Acting in Anaesthesia: Ethnographic Encounters with Patients, Practitioners and Medical Technologies
Dawn Goodwin

The Materiality of Learning: Technology and Knowledge in Educational Practice
Estrid Sørensen

How Students Come to Be, Know, and Do: A Case for a Broad View of Learning
Leslie Rupert Herrenkohl and Véronique Mertl

Making Work Visible: Ethnographically Grounded Case Studies of Work Practice
Margaret H. Szymanski and Jack Whalen, Editors

Embodied Interaction: Language and Body in the Material World
Jürgen Streeck, Charles Goodwin, and Curtis LeBaron, Editors

Games, Learning, and Society: Learning and Meaning in the Digital Age
Constance Steinkuehler, Kurt Squire, and Sasha Barab, Editors

Cultural Development of Mathematical Ideas: Papua New Guinea Studies
Geoffrey Saxe

Constructing the Self in a Digital World
Cynthia Carter Ching and Brian J. Foley, Editors

The Dynamics of Auction: Social Interaction and the Sale of Fine Art and Antiques
Christian Heath

PowerPoint, Communication, and the Knowledge Society
Hubert Knoblauch

Contested Learning in Welfare Work: A Study of Mind, Political Economy, and the Labour Process
Peter H. Sawchuk

How Humans Learn to Think Mathematically: Exploring the Three Worlds of Mathematics
David Tall

The Learning in Doing series was founded in 1987 by Roy Pea and John Seely Brown.